PRAISE FOR
GUY PERELMUTER &
PRESENT FUTURE

"A tour de force of technologies that are defining the present and questions that will shape the future, beautifully interweaved with historical narratives spanning science, fiction, and philosophy."

— **FADEL ADIB,** Associate Professor, Massachusetts Institute of Technology and Founding Director, Signal Kinetics Group - MIT Media Lab

"Gradually. Then suddenly. The cadence of technological change described in this book is paramount. Although Perelmuter covers a wide range of technology domains, the overarching theme in each case is similar. The impact of what is coming must not be underestimated. Certain decisions that may seem small today could be very consequential tomorrow. Looks can be deceiving because of the cadence—gradually, then suddenly. Every policy maker, business leader, entrepreneur, and investor should read this book."

— **AJAY AGRAWAL,** Professor, University of Toronto, Founder, Creative Destruction Lab, and Author of *Prediction Machines: The Simple Economics of Artificial Intelligence*

"Popper, in *The Poverty of Historicism*, claimed that because the growth of human knowledge cannot be predicted, the future course of human history is not foreseeable. At the same time, we like to quote William Gibson's 'The future is already here—it's just not very evenly distributed.' In *Present Future*, Perelmuter is attempting to bridge these two opposing perspectives by delving into specific technologies and using the past to forecast their future impact."

— **GAD ALLON,** Jeffrey A. Keswin Professor, Director of the Management & Technology Program, Wharton School, University of Pennsylvania

"With the context of an economic historian and the on-the-ground insights of an active technology investor, Perelmuter's *Present Future* brings readers to the bleeding edge of the science and technologies poised to revolutionize the 21st century. Comprehensive and yet enthralling, the book is a must-read for anyone who has an intellectual or commercial interest in what the future may hold."

—PETER HEBERT, Co-Founder and Managing Partner, Lux Capital

"A comprehensive survey of action across the entire frontier of advanced technologies is daunting in concept and even more so in execution. Guy Perelmuter has pulled it off, providing an accessible yet historically informed review from the world of algorithms to the world of genomic analysis by way of just about every field of science in between. Most important: he avoids the hype-ridden cheerleading that all too often accompanies accounts of breakthrough innovation. Rather his introductions to the future are thoughtfully accompanied by relevant caveats and contingencies. Thus, *Present Future* offers an informed and balanced assessment of the technologies driving our lives."

—BILL JANEWAY, Venture Capitalist, Economist, and Author of
*Doing Capitalism in the Innovation Economy: Reconfiguring the Three-Player
Game between Markets, Speculators and the State*

"Guy reminds us that the future is arriving gradually, then suddenly. *Present Future* is a fascinating, expert look at the history of the key technological advances affecting life today, and preparation for the exponential leaps yet to come. Essential reading for anyone who wants to marvel at the incredible technological advances of our recent past, and for those who want to help shape where we go from here."

—BILL MARIS, Founder and first CEO of Google Ventures,
Founder of Calico, Founder of Section 32

"In this engaging and thoughtful book, Perelmuter reminds us that the future of innovation is already part of our present, with endless possibilities. He argues that technological change—from artificial intelligence to robotics and quantum computing to nanotechnology—brings great opportunities and responsibilities. *Present Future* urges investors, entrepreneurs and governments to use the gift of innovation wisely to productively transform how society functions."

—TOM NICHOLAS, William J. Abernathy Professor of Business Administration, Harvard Business School and Author of *VC: An American History*

"As we barrel through technological change into an unstable future, it will be important to understand how innovation will touch and transform practically every sector of our daily lives. Guy has written a spotter's guide to outbreaks of the rapidly oncoming future with insight and meticulous attention to detail."

—JOSHUA SCHACHTER, Angel Investor and Entrepreneur

"Those who fight for the future live in it today. This book is an invaluable guide to how new technology will shape our future across personal, professional and public life."

—DAKIN SLOSS, Founder, Prime Movers Lab

"If knowledge is power (and it surely is), then Guy Perelmuter's excellent book *Future Present* provides readers serious power as they look to navigate an increasingly complex world. Too often books exploring the future fail to provide the kind of context the past provides. Perelmuter doesn't make that mistake. He has gathered many minds, plowed through diverse mountains of data, history and thought, and intelligently distilled it all to give us the perceptive and multifaceted handbook we all need to shape things to come with insight and intelligence."

—CHIP WALTER, Author, *Immortality, Inc.: Renegade Science, Silicon Valley Billions, and the Quest to Live Forever* and *Last Ape Standing: The Seven-Million-Year Story of How and Why We Survived*

PRESENT
FUTURE

PRESENT FUTURE

BUSINESS, SCIENCE, AND THE DEEP TECH REVOLUTION

GUY PERELMUTER

FAST
COMPANY
Press

Fast Company Press
New York, New York
www.fastcompanypress.com

This work is being published under the Fast Company Press
imprint by an exclusive arrangement with *Fast Company*.
Fast Company and the *Fast Company* logo are registered
trademarks of Mansueto Ventures, LLC. The Fast Company
Press logo is a wholly owned trademark of Mansueto
Ventures, LLC.

Distributed by River Grove Books

Design and composition by Greenleaf Book Group
Cover design by Pedro Cappeletti
Cover image used under licence from
Shutterstock.com/©Dima Zel

Translated from the original Brazilian Portuguese
by Andrea Roach

Publisher's Cataloging-in-Publication data is available.

Print ISBN: 978-1-7354245-1-4

eBook ISBN: 978-1-7354245-2-1

First Edition

For my parents,
Armand and Renée,
always present

CONTENTS

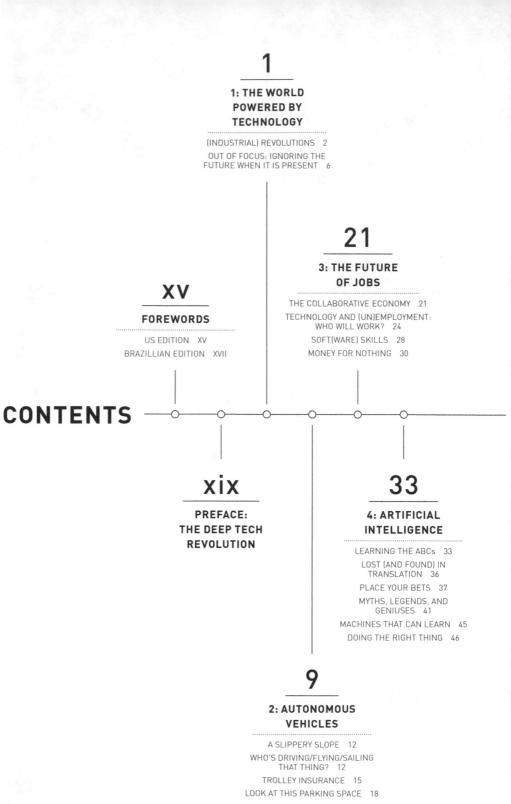

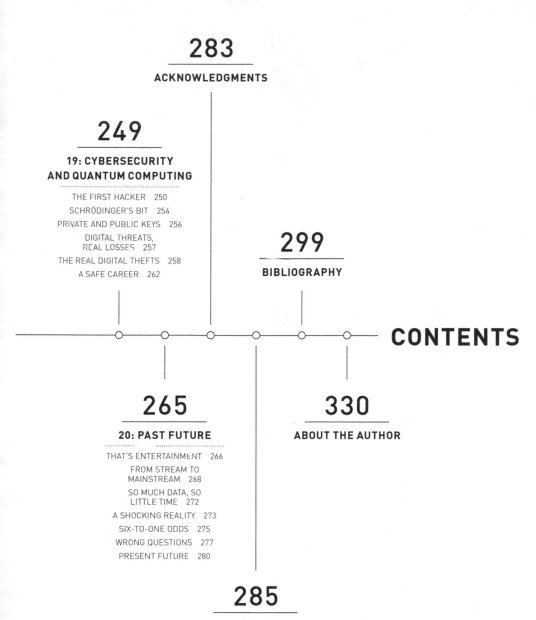

CONTENTS

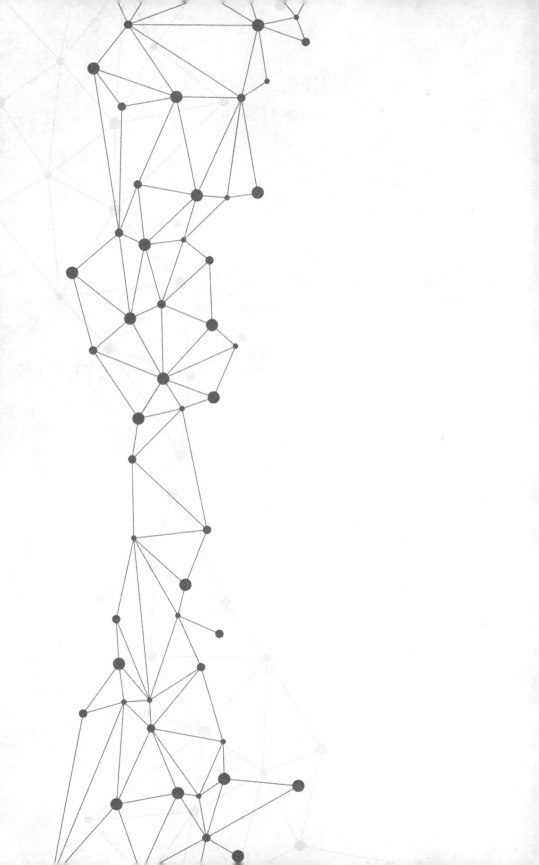

FOREWORD

TO THE US EDITION

THROUGHOUT HISTORY, EXPLORERS, PIONEERS, TRAILBLAZ-
ers, and adventurers risked everything to give us all something they never
had—a map to the present, a way forward. In discovering foreign lands,
advancing to continental cliffs and oceans' outer edges, and defying gravity to
give chase to celestial glimmers, they forged an invitation to follow them—
untethered and unhindered from ignorance and uncertainty, and equipped
with knowledge across all three dimensions of space.

But not time—for that is the dimension of historians and futurists, of
chroniclers of what was, and speculators of what may be. Here is a truth: In
making any decision, we are by definition deciding what to do . . . next. We
must choose amongst known possibilities and paths, simulate outcomes and
consequences in our minds. Another truth: At any decision point, 100% of
the information we have is based on the past, while 100% of the value and
consequences of the decision we make lies in the future, which is inherently
probabilistic and unknown. This is also the best definition of risk: that more
things may happen than will. Those who can better assess what may happen
next have an advantage over those who can't, and as a consequence they will
make better decisions, choices, and investments.

Our travels through space are always less risky and easier to navigate when
we have a critically important map of the territory.

When it comes to our endlessly unfolding future, the only certainty is
uncertainty, and the only way to reduce uncertainty is to have a deep sense of
history and reliable clues to the future.

The richest, most robust, and deeply researched histories oft come from analytical academics and obsessive historians, but the best guides to the future don't. Those clues come not from consultants or academics but from two groups: those who are inventing the future—the scientists, engineers, entrepreneurs, and founders of high-tech, high-risk ventures who have a conception of the way the world ought to be and endeavor to make it so—and those who are investing in the future.

As a fellow entrepreneur and venture capitalist obsessed with science and science fiction, with human potential to create technology, and technology's impact on human potential, Perelmuter has a valuable and rare seat alongside some of the greatest investors, inventors, and engineers of our time who are pushing boundaries and pioneering advances in artificial intelligence, robotics, satellites, biotechnology, advanced materials, energy breakthroughs, and beyond.

In *Present Future*, Perelmuter has created a deeply informed, rigorously researched, and indispensably intelligent guide to both the deep past of our technology and our near future as it unfolds. You will find this useful whether you are an inspired entrepreneur considering your next move, an investor considering where to take calculated risks, or just generally concerned, curious, or even excited to not only greet the technologies that are arriving like an alien species but to grok their social implications. It's said that the future is already here; it's just unevenly distributed. In the pages that follow, Perelmuter changes that asymmetry and gives us all who read closely a great advantage.

JOSH WOLFE, Founder and Managing Director,
Lux Capital
March 7, 2020

FOREWORD
TO THE BRAZILIAN EDITION

HUMANKIND'S STANDARD OF LIVING EVOLVED VERY SLOWLY over millennia, until the Industrial Revolution—the first one, in the second half of the eighteenth century—set in motion a spiral of extraordinary innovation that has not stopped. Since then, productivity has been evolving in a growing number of areas, which multiply and reinforce each other. This involves increasingly specialized ideas, innovations, and processes of both abstract and practical origins, which affect virtually all aspects of our lives today.

From an economic standpoint, technology reaches its full potential when it transforms abstract ideas and concepts into products and services that are born, compete, and die on the market. The person who described this process best was Austrian economist Joseph Schumpeter, who dubbed it *creative destruction*— one of the biggest drivers of growth.

Keeping pace with this awesome force is an impossible task for the vast majority of mortals—but not for Guy Perelmuter—who has given us this treasure in the form of a book. A quick glance through the contents gives us an idea of the scope of this work: autonomous vehicles, artificial intelligence, social networks, robotics, nanotechnology, big data, and much more.

A new topic is presented in each chapter, with a fascinating and beautifully illustrated backstory upon which the author builds to explain the subject, its importance, and its uses in a clear and practical way.

The application of technology to production creates opportunities and generates wealth. But it also induces fear and risk. At the top of the list of fears and risks is the future of jobs—the subject of one of the chapters. One of the biggest dreads of the contemporary world is the replacement of people by machines and systems that include artificial intelligence, robots, and other topics covered in this book. It seems increasingly clear that technology and education are natural partners. Brazil urgently needs to accelerate improvements, modernization, and access to education, or the gap between the country and the world's highest standards of living will continue to widen.

Not unrelated to technological risks is climate change resulting from global warming. There is no doubt in this case about the role of humanity in what today represents an enormous challenge to the future of our planet. This book also touches on this topic—an absolute imperative that will demand bold responses, including technological solutions, before it is too late.

It is possible to imagine genuine revolutions in areas such as health, education, finance, entertainment, transportation, socialization, and much more. This book has come at the right time. As a reader and citizen of the world, I offer my thanks to the author.

ARMINIO FRAGA, economist and former
president of the Central Bank of Brazil
October 8, 2019

PREFACE

THE DEEP TECH REVOLUTION

"THE BEST WAY TO PREDICT THE FUTURE IS TO INVENT IT."

Any of us who are curious, interested, or concerned about what lies ahead would be wise to follow this advice by American computer scientist Alan Kay, who worked at the famous Xerox PARC. Founded in 1970, it gave the world many innovations that are now part of the daily lives of billions of people, including laser printers, the Ethernet (a communication standard currently used by the vast majority of computers), and the graphical user interface (the visual elements like windows and buttons through which we interact with computers, tablets, smartphones, ATMs, and other electronic devices).

I, for one, have always been passionate about technology—not only what we have come to experience as modern technology, but pretty much any invention or device that reflects the human spirit of ingenuity and creativity. From the Greek Antikythera mechanism (probably the very first mechanical computer, used to predict astronomical events) to the quantum computers being developed today, our driving force as a species has always been a constant sense of curiosity. How do things work? Why is the universe the way it is? What can we do about it?

I was probably eight or nine years old when I first began learning how to program a computer: a Sinclair ZX81 with a whopping 64KB of (maximum!) memory and a cassette recorder storage unit. Over the next decades I was able to witness the amazing progress predicted by Moore's Law (more on that later in the book) and to realize that this very experience was nothing but one more link in a very long chain of events that was triggered when our earliest ancestors began using tools, roughly a couple of million years ago.

The very purpose of this book is twofold. First, to overthrow this myth that we are "living a period of change." The entire history of civilization is all about change—and, more than that, about *technological change*. This is what defines us as a species, this is what propels us forward. Change is coming faster and faster, that's for sure—and it will likely accelerate even more. And second, to highlight and explain not only the benefits but also the risks that a tech-driven lifestyle throws at us.

What is remarkable about the current technological changes we are experiencing is that they are sitting at the intersection of a set of extraordinary advances: faster microprocessors, cheaper digital storage, ubiquitous access to information, efficient algorithms, and an increasingly better understanding of the laws of nature. These ingredients, decades in the making, are some of the key enablers of the Deep Tech Revolution.

Deep Tech is where science meets technology, where PhDs and subject matter experts are able to apply their knowledge and transform it from intellectual achievements and academic papers into systems, devices, prototypes, products, and methodologies. Deep tech companies are the ones effectively building the future of the world economy, one technology at a time: robotics, biotech, nanotech, artificial intelligence, self-driving vehicles, energy, aerospace, agritech—the list goes on and on.

In each chapter of this book, I discuss one or more (deep) technologies that are bound to become part of our future. As you will see, I try to focus on the advances that are created to address "inevitabilities" in the making: longer human life spans; population growth; an increasing demand for energy, mobility, and food; and ever more complex systems fed by unimaginable amounts of data flowing through a vastly interconnected infrastructure over space, air, land, and sea.

I believe that understanding not only how these technologies work and what they are all about but also their remarkable origins is critical to fully understand and appreciate the magnitude of their impacts on our individual and collective futures—making sure their social and environmental impacts are not lost on us.

But when it comes to the future, we'll find no shortage of technical and scientific predictions that ended up as embarrassments. Financial services company Western Union took in revenues of more than $5.3 billion in 2019. But in an internal memo issued in 1876, 25 years after its founding, the company stated that "this 'telephone' has too many shortcomings to be seriously considered as a means of communication." In 1895, Lord Kelvin (William Thomson, 1824–1907)—who made significant contributions in the fields of thermodynamics and electricity—asserted that "heavier-than-air flying machines are impossible."

In 1903, the president of the Michigan Savings Bank advised Horace Rackham, Henry Ford's lawyer (1863–1947), not to invest in the Ford Motor Company, stating, "The horse is here to stay, but the automobile is only a novelty—a fad." Thomas Watson (1874–1956), CEO of IBM between 1914 and 1956, stated in 1943, "There is a world market for maybe five computers." Speaking about television, film producer Darryl Zanuck (1902–1979) said in 1946, "People are soon going to get tired of staring at a plywood box every night." In 1961, the commissioner of the FCC (the Federal Communications Commission, which regulates the telecommunications market in the United States) said, "There is practically no chance space communications satellites will be used to provide better telephone, telegraph, television, or radio service inside the United States."

In a 1995 *InfoWorld* article, Robert Metcalfe, cofounder of the network equipment company 3Com and one of the inventors of the Ethernet standard, wrote that "in 1996 the Internet will collapse." Two years later, in 1997, during the Sixth International World Wide Web Conference, Metcalfe literally ate his words: He placed a printed copy of what he had said in a blender with a clear liquid, mixed it all up, and drank the contents in front of the audience.

In 1965 psychologist and computer scientist J. C. R. Licklider (1915–1990)— possibly one of the greatest visionaries in the history of computing—wrote in his book *Libraries of the Future* that "people tend to overestimate what can be done in a year and to underestimate what can be done in five or ten years." And history has shown that "Lick" (as he was known) was right.

The future is already here. We're living in it. It's all around us—a *present future*—and in this book we'll take a journey to discover just what that means. We'll travel through prehistory, the ancient civilizations of the East and the West, the Middle Ages, the Renaissance, the Scientific Revolution, and the four (as of now) Industrial Revolutions. My aim is to present the explosion of new technologies that we are experiencing now as nothing more than the natural result of the work carried out over the course of history by hundreds of inventors, scientists, entrepreneurs, pioneers, and explorers. This is the nature of our civilization, but it has reached a stage at which the rate of innovation and the risks posed to our very survival are real. It is time to give them a serious look.

Let's begin our exploration of the Deep Tech Revolution and the technologies of the past, the present, and the future.

1

THE WORLD POWERED
BY TECHNOLOGY

THE WORLD EXPERIENCED WHAT WE NOW CALL THE FIRST
Industrial Revolution beginning around 1760. At that time, the production of
goods shifted from individual craftsmen to machines in factories using water
and steam power. By 1870, the Second Industrial Revolution (also known as
the Technological Revolution) popularized electricity, assembly lines, and the
division of labor. Then, once more, beginning in the 1950s, the Third Industrial
Revolution—the Digital Revolution—swept the planet, ushering in the era of
digital electronics and the so-called Information Age.

(INDUSTRIAL) REVOLUTIONS

These three powerful transitions altered the ways in which different components of production chains interacted with each other, thus impacting not only the economy, but also society, politics, philosophy, culture, and science. These revolutions shaped our world and led to unique questions and challenges for future generations.

English historian Ian Morris in his books *Why the West Rules—for Now* and *The Measure of Civilization: How Social Development Decides the Fate of Nations* presents and details a methodology that attempts to quantify the social development that has resulted from specific inventions over the course of history. In other words, how did these inventions impact individuals and society as a whole, affecting their well-being and potential?

Morris concludes that the steam engine produced the most dramatic and fast-paced effect on the progress of civilization. The acceleration of both GDP and population growth are clearly driven by the First Industrial Revolution: Few events have had as significant an impact on humanity as the change from the manual production system to the implementation of automated means of production. This change marked the beginning of a new age of social and economic development with significant changes in the way we access and distribute goods. Even more important, the rate of pollution of Earth was forever changed.

Now, less than half a century after the Third Industrial Revolution, we are seeing a new transformation that will permanently modify how we do business, produce goods, and interact with goods and people: the Fourth Industrial Revolution or what I like to call the *Deep Tech Revolution*. New technologies are enabling ideas once confined to science fiction to gradually build a more present future: integration between artificial and biological systems, learning techniques for communication between machines and their parts, and the extension of the physical reality into virtual reality (VR). The unprecedented speed and depth of this revolution stems from an auspicious confluence of factors: the increase in computer systems' processing power, the falling cost of data storage units, the decreased size of equipment and sensors, and the evolution of efficient algorithms.

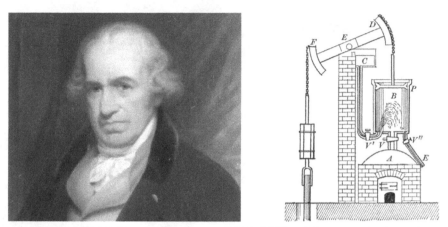

Figures 1.1 and 1.2. Thomas Newcomen (1664–1729) and his atmospheric engine.
Source: Newton Henry Black and Harvey Nathaniel Davis, *Practical Physics*,
New York (Macmillan and Company, 1913)

Figures 1.3 and 1.4. Scotland's James Watt (1736–1819) used the atmospheric
engine invented by England's Thomas Newcomen, created to remove water from
coal mines, as the basis for the development of his own steam-driven engine.
Source: National Portrait Gallery NPG 186a, Carl Frederik von Breda

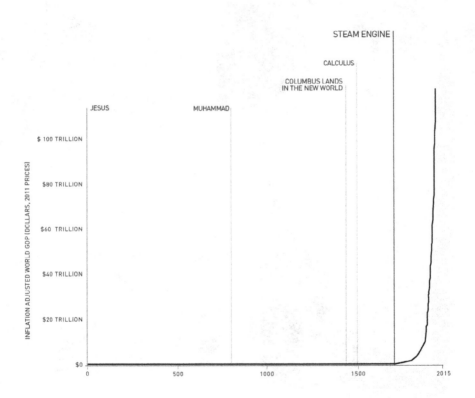

Figure 1.5. The impact of the introduction of the steam engine on humanity.
Source: ourworldindata.org based on World Bank & Maddison (2017)

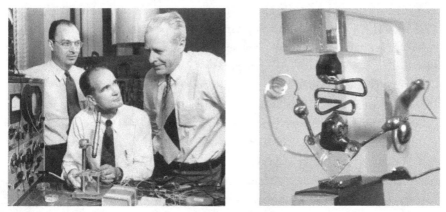

Figures 1.6 and 1.7. John Bardeen (1908–1991), William Shockley (1910–1989), and Walter Brattain (1902–1987) won the 1956 Nobel Prize in Physics for the development of the transistor, which set the stage for the age of portable electronics and microprocessors. Sources: Wikimedia Commons, Getty Images

ALGORITHMS

The crafting of algorithms—sets of instructions which, when executed, solve or complete a given problem or task—was one of the first steps toward the development of automated processes, in which a systematic approach leads to a solution. The word derives from the name of a Persian mathematician, Muhammad al-Khwarizmi (780–850), who was known in Latin as Algoritmi. He created the branch of mathematics known as algebra (in Arabic, *al-jabr*, which means "reunion of broken parts"), and he was the first to present a methodology for solving systems of equations.

How can—and how should—we prepare ourselves for this new reality? How will advances in nanotechnology, biotechnology, power generation and transmission, artificial intelligence, quantum computing, smart materials (like polymers that change their volume when exposed to an electrical current), telecommunications, robotics, the Internet of Things (IoT), 3D printing, and autonomous vehicles—just to name a few of the technologies that are currently undergoing extensive development—impact us?

"History doesn't repeat itself, but it often rhymes," said author Mark Twain (1835–1910), explaining how we can use the past to try to anticipate the future. If we can look to the past to see our future, then we will be witnessing extraordinary changes over the next few decades. From law to engineering, medicine to journalism, design to architecture, entertainment to manufacturing, and economics to education—no field of knowledge will be immune to transformations in processes, models, implementations, methods, and results.

The fact of the matter is that the future has always been present in our lives, because pretty much everything we have and live with today has once been part of someone else's vision of tomorrow. And even more new jobs, careers, companies, and empires will be created. Others will disappear or evolve into something completely different. The rate at which the world is going to experience these transformations is accelerating rapidly. It is critical that we prepare to face them—in all their forms.

What will these new technologies be? How will they impact our lives, our jobs, and our homes? How are governments, brands, industries, and services going to react? How can we leverage the opportunities that will present themselves and avoid obsolescence? The challenges we will have to face in this rapidly changing world are enormous, and no industry will get through this evolution without significant changes to their processes, products, and their very brands.

OUT OF FOCUS: IGNORING THE FUTURE WHEN IT IS PRESENT

If a picture is worth a thousand words, then a company's brand is worth ten thousand. Investopedia defines a brand as follows: "The collective impact or lasting impression from all that is seen, heard, or experienced by customers who come into contact with a company and/or its products and services." A brand can function as a statement of mission, objectives, and values. Although it is a supposedly intangible asset, it is responsible for a huge percentage of the value associated with a company.

At least three groups seek to systematically determine the financial value of the biggest global brands: Interbrand (part of marketing giant Omnicom since 1993); Brand Finance (formed in 1996 by a former officer of Interbrand); and Kantar Group (previously controlled by advertising multinational WPP, who in 2019 announced they would sell a 60% stake to Bain Capital Private Equity).

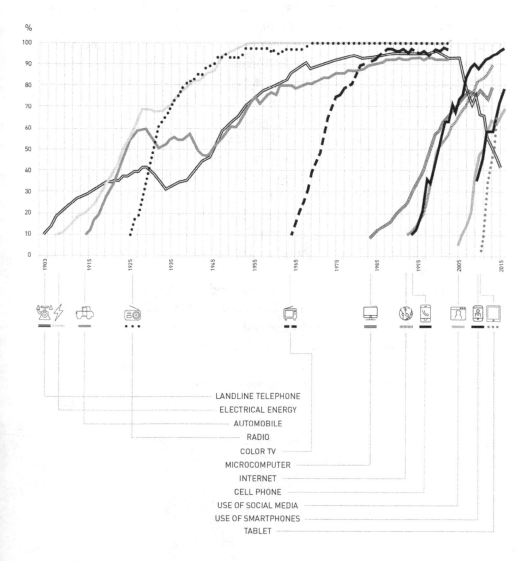

Figure 1.8. Sources: ourworldindata.org, Comin and Hobijn (2004), and others

These companies use different methodologies, but they all focus on large brands—based on criteria such as the number of licensing agreements, revenue, reach, or consumer perception. Their results contain a certain amount of subjectivity, and, consequently, there are some differences among them; nevertheless, they provide a window into the dramatic shifts that can occur. The clearest and most consistent rankings over the recent past point to the prominence of brands connected to technology—which have been around for a relatively short time in business terms: Apple (founded in 1976), Google (1998), Microsoft (1975), and Amazon (1994). When comparing these new entities to other companies such as Coca-Cola (1892), McDonald's (1940), and Disney (1923), one truth is clear: The speed, both in terms of penetration and of value creation, in the technology sector is incomparable.

A case in point is the once-valuable global brand Kodak, which was founded in 1888 and filed for bankruptcy in 2012. The advent of digital photography and Kodak's lack of response to this new technology had devastating consequences for the company. In one of the biggest ironies in the history of business, the first digital camera was actually invented at Kodak in 1975 by Steven Sasson, an engineer who was only 24 at the time. The patent obtained in 1978 earned the company billions of dollars until it expired in 2007, but executives were not willing to sacrifice a business model based on the sale of film, which had been working for decades.

Think about it: This pioneering and unprecedented technology that would have allowed Kodak to maintain its hegemony over professional and amateur photography for years to come had been developed and presented within the company itself, only to later be rejected because it challenged an old business model. In a world where the future is present, this is a hard but important lesson: Adapting to changes from new technology is imperative in the global business setting. And one of the industries whose business model will significantly change over the coming years is the automotive industry, which is over one hundred years old and includes some of the world's most valuable brands, including Toyota, Mercedes-Benz, and BMW.

AUTONOMOUS VEHICLES

THE AUTOMOTIVE SECTOR HAS BEEN AROUND FOR MORE THAN a century and is going through a time of transformation that is impacting almost every aspect of it—from the cars themselves, the roads and highways we drive on, and the cost of our insurance policies to regulatory requirements, supply chains, manufacturing processes, and as we will see in the next chapter, consumer behavior.

But let's back up. The internal combustion engine came on the scene at the end of the eighteenth century. John Barber (1734–1793) obtained the patent for a gas turbine in 1791. Over the following decades, the engine has been refined by different inventors.

OTTO AND LANGEN

German entrepreneur Nikolaus Otto (1832–1891) began his professional career selling sugar and coffee and is a prime example of how the inspiration to innovate and undertake new ventures can be sparked at any

---

continued

time in those who are alert to opportunities in the market. In 1864, he built the first functional atmospheric engine (i.e., without a compressor) with the support of Eugen Langen (1833–1895). Together, they opened the world's first engine manufacturing plant, NA Otto & Cie. In 1876, Otto patented the four-stroke engine, which is still in use, characterized by its four phases: intake, compression, power, and exhaust.

Fast forward to today, when government regulation and more environmentally conscious consumers continue to drive manufacturers toward alternative technologies (hybrid, electric, and hydrogen vehicles for land, air, and water travel). When considering investments in new energy sources, it is wise to look at current efforts to substantially improve the fuel consumption of traditional internal combustion engines and, consequentially, to reduce emissions of pollutant gases.

At the moment, electric cars only account for a fraction of the global market (approximately 1.0% in 2019), but automakers are investing heavily in models that meet consumer demands for a non-polluting car with good fuel economy and a price point comparable to existing models. In Norway, about 56% of all new cars sold in 2019 were electric, accounting for almost 11% of all vehicles in the country. The energy they use is essentially renewable because Norway's power grid is almost entirely based on hydroelectric sources. In late 2019, the IEA (International Energy Agency) estimated that China accounted for around 47% of the total number of electric cars on the world's roads, with 3.4 million units. In addition, Germany, the United Kingdom, France, the Netherlands, Israel, and India have timelines in place to permanently stop the sale of fossil fuel–driven cars over the next two decades.

Although the discussion of electric cars feels like a modern one, it is actually a return to the origins of the automotive industry, which is over 100 years old. Discoveries leading to batteries and electric engines during the nineteenth century set the stage for attentive entrepreneurs to integrate these budding businesses. Around 1890, Scottish chemist William Morrison (1859–1927), then a resident of Des Moines, Iowa, created a battery-driven six-seat vehicle that could reach a speed of just over 20 km/h (12.5 mph).

By 1895, New York City had a fleet of electric taxis traversing its busy streets, and electric cars accounted for around one-third of the cars on US roads. Electric cars were superior to those powered by steam engines (which needed water and a long time to warm up) or gas engines (which required shifting gears and were noisy and smelly). With the general acceptance of electricity, more consumers became interested in the electric car—and in 1914 Henry Ford (1863–1947) entered into a partnership with Thomas Edison (1847–1931) to develop a low-cost electric vehicle.

THE 100-MILE FRITCHLE ELECTRIC

Several different electric car manufacturers emerged, but the one who achieved the most fame was chemist and entrepreneur Oliver Fritchle (1874–1951). To prove the superiority of his batteries, which he said could go up to 150 km (95 mi) on a single charge, in 1908 Fritchle set out from Lincoln, Nebraska, toward New York in his Fritchle Victoria two seater. Arriving 20 days and roughly 3,000 km (1,865 mi) later, he dubbed his cars the "100-Mile Fritchle Electric" and opened a sales office on Fifth Avenue in New York City.

But even though cars powered by electricity had many advantages, gas-powered cars were cheaper. And in 1908, Ford introduced his famous Model T, which was half the price of electric cars, and mass production took off. That same year, US inventor Charles Kettering (1876–1958) created the electric ignition system, eliminating the need for a crank to start the internal combustion engine. (Kettering would go on to spend nearly 30 years as the head of the research division at General Motors.) Limited access to electricity outside of cities, combined with low oil prices in Texas at the time, paved the way for gasoline cars to dominate the market.

Sixty-five years later, in 1973, an oil embargo and subsequent international oil crisis brought electric cars into the spotlight again—this time as a way to lessen US dependence on foreign oil. The public had seen Apollo astronauts drive an electric vehicle on the surface of the Moon in August of 1971, but the practical results were not promising: A typical range was around 65 km (40 mi), and the maximum speed was less than 80 km/h (50 mph), so gas power held on to its position of dominance.

In 1997 the world witnessed the introduction, in Japan, of the Toyota

Prius, the first mass-produced hybrid electric vehicle. Ten or so years later, then-startup Tesla Motors produced a luxury electric car capable of traveling more than 300 km (185 mi) on a single charge. This competition led to the right market conditions for several automakers to launch their own hybrid or electric vehicles.

A SLIPPERY SLOPE

What impact will the large-scale adoption of electric cars have on the price of a barrel of oil? Assuming we keep moving toward a growing rate of adoption of electric vehicles and that the cost of batteries—a critical item in this vehicle architecture—continues to drop, to answer this question, we would need to estimate not only the savings afforded by the greater efficiency of combustion engines, but also when the number of electric vehicles will reach critical mass.

According to data from Bloomberg New Energy Finance (BloombergNEF), automotive oil consumption will decrease by two million barrels per day between 2023 and 2028 based on estimates of growth in electric car sales between 30% per year (a conservative estimate) and 60% per year (the figure seen in 2016 and 2017). This number is significant given that a similar-sized reduction in global consumption in 2014 caused a momentary drop in the price of oil of more than 50%. Still, according to BloombergNEF, by 2025 the price of battery-powered cars will be the same as for gasoline cars due to lower prices for batteries. It should be noted that the economic growth of some countries, especially emerging economies, could compensate for this potential reduction in demand for gasoline cars, and even the most aggressive estimates place the percentage of electric vehicles at 25% of all cars on the road in 2050. But, even still, these scenarios deserve attention due to their potential economic effects over the next few decades, potentially shifting prices in many commodities all over the world.

WHO'S DRIVING/FLYING/SAILING THAT THING?

We will need to prepare our societies and our economies for the introduction of autonomous vehicles with no drivers at the controls. The first autonomous vehicles came out of the work of the robotics team at Carnegie Mellon University in Pittsburgh and the Pan-European PROMETHEUS project (Programme for European Traffic of Highest Efficiency and Unprecedented

Safety) in the 1980s. Universities, automakers, and tech companies continue to work on these vehicles; Apple, Audi, BMW, Google, Intel, the Massachusetts Institute of Technology (MIT), Microsoft, Nissan, Scania, Stanford, Tesla, Toyota, Uber, and Volvo are some of the best-known organizations working in the sector.

Figure 2.1. Navlab models 1 (farthest) through 5 (front), developed at Carnegie Mellon University from 1984 through 1995. Source: Wikimedia Commons

Advances in artificial intelligence and robotics have accelerated research efforts toward an autopilot function in cars—already an old acquaintance in airplanes. Experts predict that autonomous vehicles will be the rule and not the exception over the coming decades, and, just like we receive software updates over our cell phones, the software responsible for driving your future car will also be remotely updated. This will, of course, lead to a whole new set of security-related concerns, such as the system's vulnerability to hackers. Information-security professionals—currently one of the most promising careers—will face significant challenges when cars are being driven by computer systems, as will regulatory agencies. New legislation to cover this type of driving will be needed, including safety parameters that will allow autonomous vehicles to exchange information in real time.

Autonomous vehicles use data from various sensors to determine their next action—for example, to accelerate, stop, or turn. The algorithms that make these decisions possible are trained with real world data: Unexpected events such as faulty traffic lights, unruly pedestrians, or reckless cyclists are also used for refining the algorithms responsible for driving the vehicle. However, the challenge of driving in urban environments is much more complex than the challenge of traveling on highways, moving along tracks, sailing on rivers and oceans, or flying around the world.

THE GYROSCOPE

Airplanes have been employing the autopilot function for more than a hundred years. Its inventor, American Lawrence Sperry (1892–1923), was the son of Elmer Sperry (1860–1930), founder of Sperry Gyroscope Company and co-creator (together with German Hermann Anschütz-Kaempfe (1872–1931)) of the gyrocompass, which is essential to the navigation systems of ships and airplanes. Sperry developed the system in 1912 and demonstrated it in France at the Concours de la Sécurité en Aéroplane (Airplane Safety Competition) in June 1914. The equipment effectively corrected the plane's flight path and altitude using the connection between the plane's controls and the information from the gyrocompass. Sadly, Sperry died at the age of 30 in a plane crash while flying through fog between England and France.

One infamous story involving a gyrocompass occurred in November 1916. During a flying class for a female New York socialite, both pilot and student were rescued—nude—in the harbor off Babylon, Long Island. Apparently, they had bumped the gyrocompass, which then sent incorrect information to the autopilot. According to Lawrence Sperry, the impact with the water caused both occupants of the plane to lose their clothing, an explanation that the newspapers of the time had great difficulty believing.

Around the world, airplanes' autopilot systems are called "George" by their human counterparts. The origin of this nickname is unclear, but there are a few theories. The first is simply that this was the name chosen by the pilots to refer to the new system; another is that it is an acronym for Gyro Operated Guidance Equipment (GeOrGE). And some say that the head mechanic for Sperry Corporation, who was in charge of the installation and maintenance of the equipment, was named George—and so when the autopilot was turned on, "George" was the one flying.

Figure 2.2. Lawrence Sperry (left), inventor of the autopilot for aircraft. Source: *Chicago Daily News* negatives collection, DN-007617

On ships, where maintaining a certain altitude is not an issue, automatic navigation systems use data such as speed, currents, wind direction, and wind speed to adjust course. They also use something that is part of the daily lives of all smartphone users: GPS. The Global Positioning System, developed and launched by the US Department of Defense in 1973 and approved for civilian use in the 1980s, uses a network of about 30 satellites orbiting Earth to precisely home in on and identify the location of a specific signal receiver.

TROLLEY INSURANCE

In July 2008, the US National Highway Traffic Safety Administration reported that the cause of more than 90% of road accidents is driver error. Independent studies in the United Kingdom arrived at the same conclusion, so it is reasonable to assume that this statistic is valid globally. If we remove human beings from the equation and add sufficiently robust driverless systems, the number of accidents will drop. However, even in this promising scenario, it is likely the insurance industry won't have less work to do.

Driving is an extremely complex task for a machine, and the autonomous vehicle revolution owes its existence to advances in artificial intelligence techniques, robotics, sensors, communication systems, and processing power. By nature, we are less tolerant of errors made by machines than those made by people. After all, we think, machines exist merely to serve us and should carry out their tasks efficiently and silently. Any autonomous vehicle accidents during the transition between conventional and self-driving systems will be analyzed and discussed with keen interest.

Imagine that an autonomous car detects a pedestrian crossing a road where pedestrians are prohibited. Will the car follow its embedded programming and act to protect the pedestrian? Or what happens if the system needs to choose between the life of the pedestrian or the lives of the passengers if abruptly changing course would cause a more serious accident? This kind of dilemma was discussed by American philosopher Judith Thomson in her 1976 paper "Killing, Letting Die, and The Trolley Problem." The complexity of the algorithms that control autonomous vehicles is immense since they need to reflect the multitude of situations expected to happen in the real world.

Figure 2.3. Life-and-death decisions will be made by algorithms.
Sources: Wikimedia Commons, author

How will the insurance policies be crafted for these cars? In the event of an accident, would the liability lie with the company who made the car, the onboard software, the sensors—or the vehicle owner? In October of 2015, Volvo stated that, for insurance purposes, they would be liable for any accident occurring while one of their vehicles was running in automatic driving mode. But what happens if an accident stems from a fault with the connection that enables communication among vehicles? Or what if a hacker attacks the system? Things can get complicated when it comes to insurance—one of the world's oldest industries.

When archeologists found the Code of Hammurabi in 1901 (which dates to the sixth king of Babylon from 1750 BCE), it revealed a set of laws covering concepts such as defamation, slavery, theft, and divorce. It even contained a provision stating that a loan obtained by a merchant to transport cargo over the Mediterranean Sea cannot be collected on in the event of robbery or an accident. In other words, the very definition of insurance.

In the more than 3,500 years since then, the insurance industry has naturally accrued sophistication and complexity. Experts calculate premiums using advanced techniques, and the modern world presents opportunities for drafting policies for just about everything from jewelry and horses to artwork and the weather. For it to be justifiable, an insurance policy typically needs to mitigate the risks of a chance event that could have significant financial consequences for the owner who, as a consequence, feels the need to seek protection for their asset(s).

But what happens when the probability of the risk becomes sufficiently small that the owner no longer feels the need to take out full insurance, but only partial insurance? Or when the manufacturer of the equipment in question assumes the risks that are usually transferred (via a financial transaction) to the insurance company?

The popularization of autonomous vehicles, which is widely anticipated, will make these scenarios a reality. In a 2015 study, KPMG estimated that over the next 25 years the frequency of accidents will fall by around 80% and that the financial losses caused by these accidents will decrease by at least 40%. Industry executives interviewed believe that premiums will drop, but 68% believe that their margins will not change: While accidents will be less frequent, they will be more costly because of the sophistication of the equipment involved.

Assuming that most automakers will accept liability for accidents involving their respective autonomous vehicles—which seems to be the path the industry will take—insurance companies' future clients will be precisely these automakers. Tens of millions of individuals who currently have a policy to protect their cars may only need to worry about covering events such as theft or natural disasters.

The model of interaction used by the car insurance industry is likely to be transformed. Currently, it uses a B2C (business-to-consumer) model, from the insurance company to the end consumer. But, with the emerging trend of a growing number of consumers who prefer not to own a car but rather to use one on demand via an app, the B2B (business-to-business) model between the insurance company and the automaker (or the company providing the use of the cars) seems more likely.

According to a KPMG-led study with US-based insurance executives, 42% of those interviewed indicated that they believe that the impact of autonomous vehicles on the industry will already be significant by the middle of the 2020s. This is the situation we will be witnessing as these vehicles start to replace cars with drivers. The process will not be instantaneous—there will be a long transition period during which cars with and without drivers will be traveling simultaneously on roads and highways, generating yet another layer of complexity to be addressed. Some have suggested, for example, that lanes be created on highways exclusively for use by autonomous vehicles.

But where there is innovation, there is also opportunity. Connected vehicles—that is, those equipped with technology for transmitting and receiving data via the Internet—are able to provide data in real time. If this data is used properly (and takes into account issues related to privacy), this may change the way companies charge for insurance—for the better. With this data, it is possible to analyze usage patterns, areas where the vehicle travels, parking locations, and wear of components.

During this transition toward a world dominated by autonomous vehicles, the user-based insurance model is already being offered by some providers. In this model, the amount paid by the driver is directly related to the items mentioned previously: mileage, time of use, speed, and frequency of accidents in the areas traveled, among others. Because of this, these policies are known as PAYD (pay as you drive) and PHYD (pay how you drive). The more cautious the driver, the less they pay.

LOOK AT THIS PARKING SPACE

In addition to the autonomous vehicle technical revolution, a common phenomenon among millennials—the generation born between 1981 and 1996—is the growing adoption of the culture of the collaborative economy. Rather than buying their own car, which would be parked most of the time, people are starting to think about ways to share vehicles. Zipcar, formed in 2000 and acquired in 2013 by the Avis Budget Group for $500 million; Uber, formed in 2009 and source of $14 billion in revenue in 2019; Lyft, with $3.6 billion in revenue in 2019; and other firms make cars available when and where the user wants.

With electric computer-driven vehicles that are shared rather than owned, what will our cities look like? It is not yet clear whether these changes will actually reduce the number of cars sold: Shared vehicles will travel much greater distances in their lifetimes than their individually owned counterparts and will thus undergo more wear and tear. And it is likely that at least part of the population will still own cars for trips outside densely populated urban areas.

One of the likely consequences for cities will be a reduction in the need for vast parking spaces. At the end of the workday, shared cars (autonomous or not) don't need to remain in the city and could be parked in predefined spots farther away. Freeing up valuable space promises exciting potential changes for the real estate market and the look of cities.

No matter what model ends up being adopted—autonomous vehicles or not, shared or not—charging electric cars may also provide opportunities in the real estate and infrastructure industries. It is unlikely that current gas stations will be adapted to charge cars—after all, access to electricity is practically universal. Ford, Mercedes-Benz, BMW, and Volkswagen established a partnership in 2016 for the creation of a fast-charging network around Europe with the aim of increasing the range of electric cars, and in 2017, a project rolled out that included around four hundred locations with plans for continuous expansion.

But let's take things even further: Currently, electric cars literally must be plugged in to be recharged, just like a cell phone. Once it is no longer necessary for a person to drive the car, is it conceivable that the need to plug it in to recharge will also disappear? The future of charging—not only for cars, but for all devices that use batteries—will keep moving toward wireless alternatives. Highways could have a lane that transmits power to the vehicle, and areas near traffic lights in cities may be outfitted with charging stations under the asphalt. This exciting technology, based on electromagnetic induction and magnetic resonance, is already being tested.

3

THE FUTURE OF JOBS

THE CONCEPT OF POSSESSION IS FUNDAMENTAL TO ECONOM-
ics and basic to our everyday lives: We pay for a good or an asset, and that
good or asset becomes our property. Yet this concept faced criticism from philosophers as far back as Cicero (106–43 BCE), who pointed out that private
property does not exist in natural law; it only exists in human law. Two thousand
or so years later, Englishman Ronald Coase (1910–2013), winner of the Nobel
Prize for Economics in 1991, asserted "the law of property determines who
owns something, but the market determines how it will be used."

THE COLLABORATIVE ECONOMY

The market has been moving in the direction of a new concept that relativizes the
notion of property: The shared or collaborative economy can be defined as "a setting in which people or companies connect via online tools to offer resources of
any nature to parties who are interested in using them."

Younger generations, seeking efficiency, minimalism, or convenience, obtain
services when the need arises. Need a place to live for a few days? A temporary

place to work, preferably collaborative? A car to visit a friend? A dress made by a famous stylist or a chic handbag for a special occasion? Painting, carpentry, general chores? Private classes? All you need is a smartphone.

Instead of *shared economy*, we might better use *access economy*—most of the types of businesses listed previously don't own assets—they simply broker users' access to them. Airbnb, for example, does not own the properties it facilitates access to, but when it raised funds in March 2017, it became a company (at least on paper) valued at around $30 billion. On the other hand, the Hilton chain of hotels, in existence for one hundred years as of 2019, owns hundreds of hotels around the globe and closed out 2018 with a market value of approximately $21 billion. While traditional businesses handle all the expenses and logistics as part of their operations, Airbnb simply offers access and the interface with the client. It is no surprise, then, that in April 2016 the Accor chain acquired Onefinestay—a luxury version of Airbnb.

Facebook and Google do the same profitable thing: connect the consumer with services or products. The costs and infrastructure to create these products and services are significant, but this does not have any bearing on the interfaces. The big winners of the shared economy (or the access economy) are the owners of these interfaces. The more elegant, simple, and efficient, the better.

But the opinion on this shared, collaborative, or access economy—whichever name we choose—is far from unanimous. Its proponents point to benefits such as the reduction of waste (fewer units per person) and the increase in consumer independence (when I want it and where I want it); they say that the consumer will only use what they need (little to no inventory). And if they are not satisfied with the service, consumers can switch providers simply by pressing a button. But that's only part of the story.

At present, the owners of the interfaces let users rank, sort, and reward service providers since consumers are more interested in the opinions of their peers than those of critics or experts. Many providers believe the best way to provide reliable recommendations to their users is by looking at the pattern of consumption associated with each individual.

Netflix put out a challenge in October of 2006 offering a million dollars to anyone who could produce a better algorithm to recommend movies that a given person might like based on ratings that user had given to other movies. The competition ended in September 2009 with a winner who achieved results approximately 10% better than Netflix's original algorithm. But technological advance is relentless. During the time of the challenge, the company's business model changed: DVD rentals gave way to Internet-based transmission of videos (streaming). The winning algorithm, by then out of sync with the new model, did not end up being fully incorporated into the platform.

One fact is clear. For practically all businesses, competition takes place in the online marketplace—the place where consumers will seek their desired product or service, and where opinions and critiques that are immediately shared will determine the success or failure of the experience. Car rentals, hotels, fashion, services, dining, tourism—the more social the experience, the greater the enthusiasm and passion a community will display.

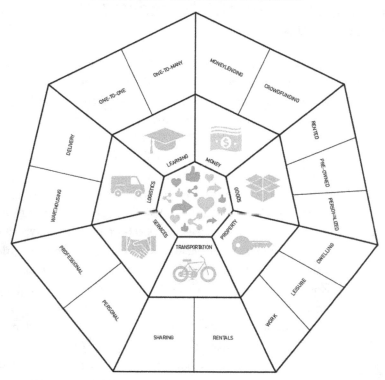

THE DYNAMICS OF THE COLLABORATIVE ECONOMY

Figure 3.1. Sources: Author, additional input based on J. Owyang and V. Mirkovic

Why would someone open up their house to a stranger? Or use part of their time driving their own car hoping for customers? Why would someone prefer to give up a job with benefits and paid vacations for something temporary that often lacks a traditional employment relationship? How will governments react? What will happen to individuals and society over the short and long term? Which careers will flourish, and which will be destroyed by automation?

The answers to these questions are likely related to a tougher, more complex, and competitive labor market—something set to become even more pronounced in the future.

TECHNOLOGY AND (UN)EMPLOYMENT: WHO WILL WORK?

Will technology improve or impair our employment prospects? How will young people fare? Finding the answers to these questions is a complex and ongoing process. Throughout the entire history of civilization, new technologies have precipitated a range of responses in the workplace.

The Greek philosopher Aristotle (384–322 BCE) wrote that "servants are an instrument that should be prioritized over all other instruments," and that if there were a way to carry out a given task without human interference, that way should be selected, thus freeing up people for other activities. The governments of some ancient civilizations (like Pericles's [495–429 BCE] in Greece) sought ways to occupy populations who became unemployed due to technical innovation; some went to the extreme of rejecting or even banning any innovation that would impact the labor market. According to economist and historian Robert Heilbroner (1919–2005), during the Middle Ages people who tried to trade or promote merchandise that could be classed as an innovation were executed as the worst types of criminals. In his 1953 book, *The Worldly Philosophers: The Lives, Times, and Ideas of the Great Economic Thinkers*, he tells the story of 58 people literally broken using a medieval instrument of torture known as the *Catherine wheel* because they sold forbidden goods.

The Luddite Movement took place in England during the First Industrial Revolution among workers who saw their labor being replaced by machines; it also inspired today's Neo-Luddism, a philosophy that is basically opposed to technological development. Some say that the word *sabotage* comes from the word *sabot*—wooden shoes that laborers at the end of the eighteenth century and the start of the nineteenth century would throw into industrial machinery to damage it.

Figure 3.2. The Leader of the Luddites. Engraving from 1812.
Source: Working Class Movement Library catalogue

When modern economic science was born, a practically answerless debate also began: Does technological unemployment, that is, the shortage of jobs caused by the substitution of human labor by machines, exist? Thomas Malthus (1766–1834) and Karl Marx (1818–1883) argued that it does, but Charles Babbage (1791–1871), one of the most important figures in the history of computing, and Jean-Baptiste Say (1767–1832, "supply creates its own demand") said that it does not. The discussion continued into the following century, when the evidence at the time pointed to a positive outlook for the future, in spite of two world wars. The general opinion was that technological progress was improving the quality of life of workers and their employers.

JEAN-BAPTISTE SAY
ECONOMIST
1767–1832

CHARLES BABBAGE
SCIENTIST,
MATHEMATICIAN,
PHILOSOPHER,
ENGINEER, AND
INVENTOR
1791–1871

IS TECHNOLOGICAL UNEMPLOYMENT REAL?

THOMAS R. MALTHUS
ECONOMIST
1766–1834

KARL MARX
PHILOSOPHER,
SOCIOLOGIST,
AND JOURNALIST
1818–1883

Figure 3.3. A few famous parties to the discussion on technological unemployment.
Sources: Cité de l'économie et de la monnaie; Wikimedia Commons, PD-US

In the final years of the last century many thinkers, economists, and journalists pondered the medium- and long-term effects of globalization, and once again innovation and automation and their potential impacts on the labor market became the focus of debate. In 1996, two European journalists, Hans-Peter Martin and Harald Schumann, published *The Global Trap*, arguing that only 20% of the economically active population would be enough to keep the world economy moving, forcing governments to support the other 80%. The 1995 book *The End of Work* by US economist Jeremy Rifkin also anticipated the elimination of millions of jobs due to technological innovation and the growth of the government-supported voluntary services sector.

Up to now, innovation has been the catalyst for so-called creative destruction, that is, jobs are not eliminated, but rather are transferred to other sectors (for example, from the agricultural sector—which has been mechanized to the extreme—to the services sector). But there are those who think this scenario is about to change—for the worse.

Historically, progress and innovation boosted the quality of life of several layers of the population, and typically only jobs that required lesser qualifications were eliminated by machines. The onset of the Fourth Industrial Revolution has intensified the discussion around technological unemployment given that a broad set of new technologies (robotics, artificial intelligence, and 3D printing, among others) has simultaneously reached a large number of industries and businesses.

According to the United Nations, the world population rose from under a billion in 1800 to 7.7 billion in 2019, while the percentage of the population living in urban settings rose from 3% to 55%. The mechanization and modernization of agricultural activity sent a significant portion of the labor force from the fields to the cities—less than one-third of the world's labor force is in the fields, and in developed countries this figure is less than 5%. Innovation has transferred jobs to other sectors—like manufacturing and services—through creative destruction.

Many believe that the rejection of globalization may be in large measure attributed to the fact that the middle class feels threatened in a world that is more connected and technology dependent, especially in light of the reduction of assembly-line jobs and the associated boost in productivity, which are accompanied by increased unemployment.

In March 2015, through the Centre for Economic Policy Research, Georg Graetz and Guy Michaels published a paper analyzing the economic impact of industrial robots in 17 countries over 15 years. The productivity and growth of the countries increased, and the hours worked by humans dropped.

In 2013, Carl Frey and Michael Osborne, of Oxford University, published a paper on the future of employment wherein they analyzed the likelihood of automation for 702 types of occupations and then overlaid the results onto the US labor market. According to them, no less than 47% of jobs were at risk of automation—and the lower the salary paid and level of education required to perform those tasks, the greater the probability of the job being replaced by artificial labor. Think about that: Nearly half of the activities analyzed showed up as being susceptible to automation.

Using different criteria not based on professions but on the subtasks performed by workers, researchers Melanie Arntz, Terry Gregory, and Ulrich Zierahn from the Center for European Economic Research (*Zentrum für Europäische Wirtschaftsforschung*) based in Mannheim, Germany, arrived at a very different conclusion: Rather than 47%, their estimate is that only 9% of the professions studied run a high risk of being automated. Other studies published by global consultancies have produced estimates of this figure between 30% and 50%.

SOFT(WARE) SKILLS

Despite the difference of opinions on the effects of technological progress on jobs, the prevalent academic and economic outlooks were that the results of technological evolution would be favorable (until recently). We used to think technology would ultimately generate more jobs, but this outlook seems to be changing. A significant number of economists and academics are starting to believe that we are entering a period where not only the number of jobs will decrease, but the type of decrease will change: It won't be just jobs that require repetitive activities or less education.

Larry Summers, former US Treasury secretary and Harvard professor, summarized this viewpoint by asserting that the future concern of humanity will no longer center on producing enough goods, but rather on being able to create enough jobs for everyone. The forces that impacted the agricultural and manufacturing sectors are now affecting services in a process that is likely to intensify. Drivers may no longer have work due to autonomous vehicles. Businesses may not need cashiers, because consumers could be charged as they exit the store by means of integrated computer vision systems and sensors that will scan their items. In fact, consumers may have already been identified when they enter the shop via automated image-recognition systems.

When we call customer service or interact over chat with a support agent, it is likely to be a machine on the other end of the line. Using artificial intelligence techniques, these *bots* start by learning how to deal with the most common requests, and, over time, accumulate experience and flexibility. There are several cases where these entities are already able to interact with humans without us noticing that we are actually communicating with a machine: At Google's I/O Conference held in 2018, CEO Sundar Pichai presented a demo of the company's intelligent assistant scheduling a haircut appointment and making a restaurant reservation.

Robotic process automation (RPA) is likely to grow significantly in the coming years (according to the consultancy Gartner, revenues for the sector totaled around $850 million in 2018). This automation means that repetitive tasks (such as copying values from a PDF file into a company's accounting system) are executed by bots, which work seamlessly in the background unless a problem that needs to be reviewed by a human is found.

Technological developments, and in particular techniques connected to artificial intelligence and data analysis, have been generating innovations that promise to substantially alter the profile of the labor market in the coming decades. The use of subjective judgment, emotional intelligence, and adaptability to unexpected situations are emerging as important characteristics for the employees of the future since these are features that are quite uniquely human and will very likely not be replaced by a machine in the foreseeable future. Jobs that had previously been seen as untouchable are now becoming endangered species as a result of these changes: translation of texts, preparation of contracts, image analysis, accounting, and financial advising, just to name a few. And there is no doubt that much more is on the way—including new careers that simply don't exist yet or that have not yet become relevant—as technology creates the need for new tasks and unexpected, promising specializations.

In his 2015 book *Rise of the Robots,* American author Martin Ford argued that the impact of the automation of tasks—both the simple and the complex—has the potential for creating a deflationary spiral: With the reduction in size of the labor market, consumers will feel less secure about spending. Several economists and a few players in the technology industry think the solution would have to be government programs that guarantee a minimum income for large portions of the population—a basic universal income.

MONEY FOR NOTHING

In 2014, Erik Brynjolfsson and Andrew McAfee, two researchers from MIT, wrote *The Second Machine Age: Work, Progress, and Prosperity in a Time of Brilliant Technologies*. They postulated that increasingly sophisticated tasks—like manipulating complex physical objects, translating texts, and approving or rejecting mortgages—may end up being carried out by machines. It is not yet possible to assess whether this type of automation will in fact eliminate jobs or simply create a change in the profile of the labor market (as has been the case over the course of history).

English philosopher Thomas More (1478–1535), author of the book *Utopia* (1516), and Spanish humanist Juan Luis Vives (1493–1540), a friend of More's and considered to be one of the fathers of modern psychology, were possibly the first to document and structure the idea of a minimum basic income: paying all citizens from a certain place an annual income, independent of any other factor. Its proponents argue that this would allow for more generalized access to education and innovation; it would make money available for parents to invest in their children; it would make it possible for people to seek out fulfillment, even in unpaid jobs; and it would offer a broader and more widespread perception of social justice. In other words, it would provide an improvement to the quality of life and to the general well-being of society. Several of the human needs discussed by the American psychologist Abraham Maslow (1908–1970) are addressed by basic income policies (please refer to Figure 3.4).

Programs exist around the world that have sought or that seek to, at least in part, implement the payment of a minimum income to a portion of the population (sometimes regardless of employment status): the United States, Canada, Namibia, Uganda, India, Mexico, and Brazil can produce statistics around their assistance programs (such as the 2008 Namibian Basic Income Grant Coalition and the Ontario 2017 Basic Income Pilot). The evaluation of these programs' effectiveness is not simple, and it typically induces arguments both for and against them. In 2010, for example, the German Parliamentary Petitions Committee heard Susanne Wiest, a defender of the concept of a universal basic income, and rejected her proposal as unfeasible for various reasons: lack of motivation for people to work, the cost of implementation, a consequential rise in immigration, and an impact on the prices of consumer goods to cover the costs of the payments.

MASLOW'S HIERARCHY
OF NEEDS AND THE EFFECTS
OF A UNIVERSAL BASIC INCOME

NEEDS

EFFECTS

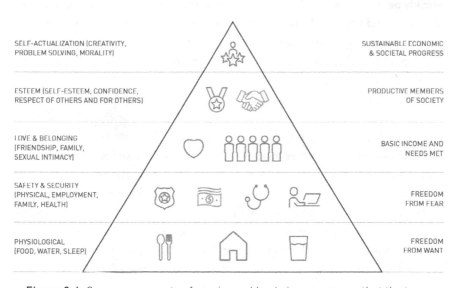

SELF-ACTUALIZATION (CREATIVITY,
PROBLEM SOLVING, MORALITY)

SUSTAINABLE ECONOMIC
& SOCIETAL PROGRESS

ESTEEM (SELF-ESTEEM, CONFIDENCE,
RESPECT OF OTHERS AND FOR OTHERS)

PRODUCTIVE MEMBERS
OF SOCIETY

LOVE & BELONGING
(FRIENDSHIP, FAMILY,
SEXUAL INTIMACY)

BASIC INCOME AND
NEEDS MET

SAFETY & SECURITY
(PHYSICAL, EMPLOYMENT,
FAMILY, HEALTH)

FREEDOM
FROM FEAR

PHYSIOLOGICAL
(FOOD, WATER, SLEEP)

FREEDOM
FROM WANT

Figure 3.4. Some proponents of a universal basic income argue that the two bottom layers of the Hierarchy of Needs, developed in 1943 by American psychologist Abraham Maslow (1908–1970), are addressed by this policy. Source: Author, based on www.reddit.com/r/economy/comments/6dcr5d/ basic_income_maslows_hierarchy_and_effect_on/

In June 2016, Switzerland held the world's first referendum on the payment of a monthly income of approximately $2,500 to its inhabitants—and the proposal was rejected by 76.9% of voters, possibly because it did not include the elimination of other government assistance programs. In January 2017, the Finnish government began paying €560 per month to two thousand unemployed persons and had plans to expand the program to people who were employed. But nearly two years later the experiment was discontinued after losing popular support when taxpayers were informed that they would need to pay more to keep the program running. In spite of this, the payment model for a minimum basic income continues to be discussed in countries such as Scotland, Spain, the Netherlands, France, and Canada (which implemented a minimum-income program in the province of Manitoba in the 1970s).

Extreme caution should be taken when considering the long-term effects that minimum-income programs could have on individuals and on society as a whole. In addition to the potential cost that programs of this nature could incur on the budget of any country, is there a risk that a good portion of beneficiaries may simply opt out of any productive activity since their subsistence would already be guaranteed? Would a program of this type not pose a risk to education and productivity? In fact, there are many places where it is apparent that people with a lower level of education who enter the program do effectively stop looking for work and don't use the money they receive to seek further academic or professional qualifications.

Looking at the past and at the way innovation has affected the labor market and people's quality of life, future prospects appear positive. On one hand, jobs have been eliminated or replaced—from agriculture to manufacturing to services—but on the other hand, several new careers have been created due to innovation and progress, and the benefits generated by the new technologies have been enjoyed by a large portion of the population. Despite the concern of a growing number of economists about the effect of technological unemployment on careers that require higher education, it still seems premature to imagine that the world will simply stop offering work. New tasks, new processes, and new possibilities will be, in and of themselves, an engine of growth for future newcomers to the marketplace.

4

ARTIFICIAL INTELLIGENCE

WHAT DISTINGUISHES A CAREER THREATENED BY ROBOTS, computers, or other devices from one that is relatively protected from the risk of automation? Which careers are immune, and which ones are doomed? It used to seem that the most severely threatened jobs were those associated with repetitive, manual tasks, and these positions have effectively been undergoing automation for a while now. As we saw, the more predictable the task the greater the chance that an artificial entity will be capable of executing it—and now, tasks that require some form of logical reasoning are also being automated.

LEARNING THE ABCs

Advances in artificial intelligence allow computers to infer and extrapolate based on situations presented and taught over time. The analysis techniques for large quantities of data (in a field popularly known as *big data*) are exponentially leveraging machines' capacity to process information and generate recommendations. As algorithms become more efficient, machines execute tasks that could once only be done by humans—like translating texts, driving cars, writing legal contracts, and recognizing complex images, just to name a few.

Now, the effects of Moore's Law (conceived by Intel cofounder Gordon Moore in 1965) can be added to this scenario. He predicted that computer processing capacity would double every two years—and this is, in fact, what we have been seeing (at least approximately) over the past five decades. (Whether this trend will continue in the future is still the subject of debate, but it seems reasonable to assume the trend toward more powerful processors and more efficient computer architectures will persist.) New development techniques for integrated circuits (such as quantum computing, increasingly cheap data storage, algorithms capable of digesting unimaginable volumes of information in ever-shorter time frames, faster processors, and artificial intelligence techniques that ensure flexibility and adaptability) are readily available ingredients that make machines capable of solving tasks normally associated with humans.

NUMBER OF TRANSISTORS
IN INTEGRATED CIRCUITS

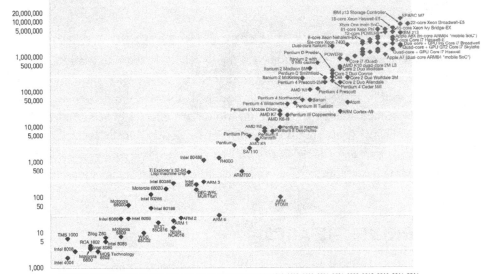

Figure 4.1. Practical illustration of Moore's Law.
Sources: en.wikipedia.org/wiki/Transistor_count; ourworldindata.org; Max Roser

We know that historically the impact of automation on productivity has tended to be positive. The McKinsey Global Institute reported that this positive influence was observed after the introduction of steam engines during the First Industrial Revolution, which led to gains of 0.3% per year between 1865 and 1910, and after the adoption of the first industrial robots and the large-scale use of information technology. The latter two events created annualized gains of 0.4% and 0.6%, respectively, between the mid-1990s and 2000. This study showed that the anticipated global growth from productivity improvements is likely to be somewhere between 0.8% and 1.4% per year over the next 50 years.

However, an increase in productivity is not the same thing as an increase in jobs. Let's take the processes of identifying, grouping, and forwarding letters, envelopes, and packages around the world. Correctly reading the recipient's address is the most critical step in ensuring the mail gets to the right destination. Nowadays, reading the address is no longer done by humans; it is done by machines. Artificial vision systems have been trained on a universe of hundreds of thousands of examples of different types of handwriting and fonts. After learning how many ways the letter *t* can be written—slanted or upright, with the cross higher up or in the middle, in print or script—the systems are able to correctly read what the user has written in practically 100% of cases. When an error occurs, the context in which the letter appears (i.e., the letters coming before and after it) helps to correct the problem for the artificial system. And if the error still cannot be resolved, it is likely that a human operator would have the same problem.

Reading is an excellent example of a task not normally associated with machines that has been mastered by technology. And the challenges are getting even more interesting: Scientists and researchers are working on artificial intelligence systems that are able to understand the meaning of a word, a sentence, or a paragraph. Computer science specialists and linguists are working together to address the challenges of contextualizing the nuances that our brains have no difficulty in understanding (such as irony, sarcasm, metaphors, or analogies).

By coupling this with a voice recognition system (available on smartphones and devices such as Google Home or Amazon Echo), it's easy to see why you are unlikely to be greeted by a human when you call a customer support line or chat with a support analyst. You'll more often be interacting with bots—software applications built to serve people in the most natural way possible, with no wait time. Bots have unlimited patience and are available 24/7.

This type of service is invaluable. Forrester Research's 2016 report on market trends for client support services stated that no less than 73% of people said that the most important thing a company can do to offer good service is to value their clients' time.

It is, therefore, interesting to note that nearly 30 years ago, American inventor Hugh Loebner founded the Loebner Prize aimed at promoting research in artificial intelligence. To determine the winner, judges interact, via chat, with either a human or artificial counterpart. The chatbot that is taken to be a human by the largest number of judges is named the winner. While there is no consensus on the matter—some professors don't consider the contest to be very scientific—the interest and curiosity it sparked illustrates the advance in machines' capacity to interact with humans.

LOST (AND FOUND) IN TRANSLATION

Voice is one of the most intuitive forms of interaction. We know machines have the capacity to read text or to hear a sound, but machines can also understand what a person is actually saying. The widespread adoption of voice-driven devices in our homes is just a matter of time—and the major tech companies have already given identities to their digital assistants: Amazon has the Echo (a device connected to the personal assistant Alexa), and Google has Google Home. Apple has loaded Siri onto their cell phones, Microsoft has created Cortana for the Windows environment, and Google has developed the Google Assistant for the Android system. Beyond convenience, the potential for voice-driven applications will expand, allowing seniors, people with disabilities, and children who can't yet read to interact with the devices.

Another technology for interaction between humans and machines involves the interpretation of signals generated in the brain. In September 2019, Facebook acquired the startup CTRL-Labs (for an undisclosed amount). The firm is developing a wristband that captures and transmits the nerve impulses responsible for hand movements, which makes it possible for someone to control a smartphone, tablet, or other electronic device simply with a thought.

Understanding simple voice commands—"turn on the lights" or "play music"—is one thing. In general, the more restricted the vocabulary, the simpler the task is for the computer. The same logic applies to translation—one of the tasks that is increasingly being handled with the help of smart systems. For example, when a cellphone camera is pointed at a text written in a foreign language, the text is translated into the language selected via Google Translate. The results are not always accurate, but in most cases the translation is reasonably understandable.

So-called machine translation marries information technology and linguistics to translate oral or written content from one language into another. More than simply replacing each word from language *A* with the same word in language *B*, machine translation's objective is to correctly interpret the *meaning* of the original text—something much more complex and sophisticated. One of the ways of studying this complexity is through what has become known as the Winograd Schema Challenge, inspired by the work of Stanford University computer science professor Terry Winograd. The challenge is to create a machine that can correctly interpret ambiguous sentences—a task that humans normally manage to do fairly easily but is extremely difficult for algorithms. Winograd also supervised Larry Page—who suspended his PhD studies in 1998 to found Google with Sergey Brin, another PhD candidate.

Microsoft, which acquired Skype's software for voice and video calls for $8.5 billion in 2011, used machine translation techniques to develop a translation module for text and for voice. This means you can call someone who doesn't speak your language and still be able to communicate without the need for a translator—the software performs this task in real time. The service works for a few languages in voice/video mode and for a long list of languages in chat mode. Google, via integration with the Google Translate service, is also headed in the same direction. Improvements are needed, but constant refinement is ongoing, and at some point in the next few decades the language barrier will no longer be an impediment to people's direct communication.

Just like other equipment that responds to voice commands, the narrower the vocabulary used, the better the results. Technical texts can already be translated quite satisfactorily, but the same cannot be said for literary texts in general—and questions remain as to whether a machine can master the subtleties of non-technical translation and the interpretation of the immense range of human emotions.

PLACE YOUR BETS

The International Labor Organization, an agency of the United Nations, compiles data on the world labor market that shows the dominance of the services sector over the agricultural and manufacturing sectors in terms of the number of people employed. It is estimated that in 2019 around 50% of the worldwide labor market was engaged in activities in the services sector. But if only the economies of OECD (Organization for Economic Cooperation and Development) member countries are considered, this number jumps to 73%, or almost three of every four workers. In countries with the lowest socioeconomic development indexes as reported by the United Nations, nearly 55% of the population works in the agricultural sector (in OECD countries, this number is approximately 5%) and 32% in the services sector.

In developed economies, where the cost of labor is higher and it makes undeniable economic sense to automate processes and methods, innovation has substantially reduced the need for labor in the industrial sector. The percentage of humans working in plants and manufacturing in these countries has been dropping significantly, and in 2019 the International Labor Organization estimated that no less than 79% of employed Americans worked in the services sector.

Looking at the data, it is reasonable to imagine that the reduction in jobs due to technical and scientific advances that increased efficiency, accuracy, and safety in the agricultural and industrial sectors will also occur in the services sector. We've discussed machines' capacity to read, and we've discussed advances in the area of understanding, using the example of document translation. The next step is extrapolation—or the ability to draw conclusions from the analysis of a given data set.

The financial market produces vast quantities of information every day. The market's reaction to news stories is instantly reflected in the movement of stock prices. A natural disaster, the results of an election, or the release of inflation rates all impact the prices of assets in different ways. Artificial systems are already being used to deduce how a given company will respond to specific events by analyzing historical patterns, correlations among assets, and other parameters—while being able to answer questions crafted in natural language—and to prepare reports without human supervision.

These reports are typically produced by NLG (natural-language generation) systems. The architecture for these systems is relatively simple: The input data is structured information, such as charts and tables. The finished product is a descriptive report—a text produced completely autonomously, containing the most relevant information from this data. Banks, credit card companies, news agencies, and technology firms already use this technology for non-repetitive tasks that require an understanding of the context and the ability to process a large number of variables. Computers are becoming so good at this

DISTRIBUTION OF JOBS PER SECTOR

2019 – WORLD

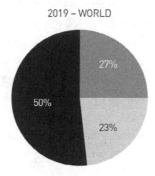

2019 – LEAST DEVELOPED COUNTRIES (UNITED NATIONS)

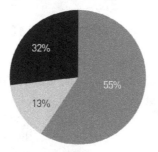

2019 – DEVELOPED COUNTRIES (OECD)

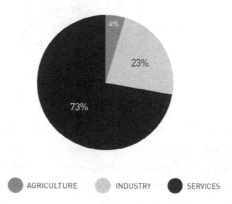

AGRICULTURE   INDUSTRY   SERVICES

Figure 4.2. Copyright © International Labour Organization 2019

task that they can now outperform their human counterparts when it comes to achieving goals that depend on knowledge and cognitive capacity—that is, winning at games.

In the world of electronic games—a $150 billion industry as of 2019—the presence of an adversary controlled by a computer is nothing new. Beginning with the first games, such as *Speed Race* (1974) and *Space Invaders* (1978), human players have been accustomed to facing opponents controlled by machines. The gaming industry quickly evolved and developed new programming techniques and used more processing power as it became available. Games that had typically been reserved for humans, such as checkers, chess, and backgammon, began to appear on the list of one-person games. There is an important difference between the latter and the former: The computer's behavior depends on the actions of the human.

THE SHANNON NUMBER

In a 1950 article, American mathematician Claude Shannon (1916–2001)—one of the most important figures in the history of computers—estimated that in a typical game of chess, there are at least 10^{120} possibilities for different configurations (this number has since become known as the Shannon number). To illustrate how big this number is, the current estimate for the number of atoms in the entire observable universe is around 10^{80}. Thus, a solution that seeks to test all possibilities for an entire match would be unfeasible—so as the player makes each move, the algorithm simulates the next moves. The number of simulations depends on the complexity of the program and on the processing and storage power of the computer.

Games involve possibilities, strategies, bluffs, and intuition that would be difficult to reproduce using traditional programming techniques. With the advances in processing, storage, and coding capacity; new programming models; and higher quality graphics cards, the sophistication levels of new games and player experiences have increased notably.

The extrapolation capacity acquired by machines that learn by using computational intelligence techniques, statistics, and pattern recognition has enabled companies such as IBM, Google, and Microsoft to build their own cognitive systems that are able to emulate a fundamental human characteristic that is the key to the decision-making process: reasoning. One of the most effective ways of demonstrating the reasoning capacity of these systems was to make them perform activities that humans understand well and that are not repetitive.

In 1996 and 1997, Deep Blue (an IBM computer) defeated the then-world chess champion, Garry Kasparov. In 2011, Watson, also by IBM, beat two champions of *Jeopardy!* One of the most extraordinary feats of artificial intelligence was Google DeepMind's victory over Lee Sedol, a South Korean professional player of *Go*, a more than 2,500-year-old Chinese strategy game. The story was turned into a 2017 documentary, *AlphaGo*, directed by Greg Kohs. It does seem that in a deep tech world, *Deep* will be a popular first name for new technologies.

Figures 4.3 and 4.4. Human versus machine: in 1996, a computer defeats the world's chess champion; twenty years later, it masters the complex game of Go. Source: Getty Images

These victories are iconic because they showed the world that artificial entities could operate in domains that require faculties associated with humans. These are *thinking machines*, which spark both curiosity and fear and inspire books and movies about an apocalypse caused by robots and computers.

MYTHS, LEGENDS, AND GENIUSES

The conflict between humans and machines is an ongoing one, and it is closely connected to deeper, more philosophical matters related to the concept of artificial intelligence: *cogito, ergo sum*—I think, therefore I am. In his 1637 work,

Discourse on the Method, philosopher and mathematician René Descartes (1596–1650) established in a simple and elegant manner the relationship between the ability to think and the conviction of an absolute, true, and unquestionable existence. The act of thinking is what sets humans apart from all other living creatures on the planet—we are able to abstract, doubt, and question.

Since ancient Greece, thinking machines—humanoids or not—appear in myths and legends. *The Iliad*, attributed to Homer, talks about robots made by the Greek god of artisans, Hephaestus (or Vulcan, according to Roman mythology), and Chinese legends from the same time period mention machines endowed with intelligence. Throughout our history, humanity has always pondered the possibility of transferring the capacity of thinking, and, therefore, of existing, to inanimate creatures.

One of the most famous examples is the 1818 work by English writer Mary Shelley (1797–1851), about a scientist named Victor Frankenstein who gives life to (and later rejects) a monstrous creature made from inanimate parts. Shelley was married to Percy Bysshe Shelley (1792–1822), an English poet and friend of the notorious Lord Byron (1788–1824), whose only legitimate daughter was computer science pioneer Ada Lovelace (1815–1852).

Between 1930 and 1940, Alan Turing (1912–1954), a central figure in the history of computer science, mathematically established the fundamental concepts for the development of modern computers and the area of artificial intelligence. Turing proved that a binary system—made up of only two symbols, such as 0 and 1—would be able to solve any problem as long as it was possible to represent it using an algorithm. The so-called Turing machine is a powerful tool, and its formalization in Turing's 1936 article *On Computable Numbers, with an Application to the Entscheidungsproblem* (or "decision problem") remains one of the most important scientific works in the theory of computation.

And as if this wasn't enough, Turing also played a critical role during the Second World War (1939–1945), working in the intelligence and security organization known as the United Kingdom's Government Communications Headquarters, or GCHQ. His work was the subject of the 2014 film *The Imitation Game*, directed by Norway's Morten Tyldum and based on the book by English author Andrew Hodges, *Alan Turing: The Enigma*, published in 1983. The title refers not only to Turing himself, but also to the encryption machine developed by German engineer Arthur Scherbius (1878–1929) and used by Nazi Germany during the war. Cryptoanalysts led by Turing at GCHQ at Bletchley Park cracked the German code—and, according to several experts on the topic, this breakthrough reversed the direction of the war.

Figures 4.5 and 4.6. LEFT: Ada Lovelace (1815–1852), English mathematician and the first person to publish a program to be processed
by Charles Babbage's analytical machine (see Chapter 3). RIGHT:
Alan Turing (1912–1954) at Bletchley Park, considered by many to be
the father of computer science and of artificial intelligence. Sources:
Getty Images (PD-US), Wikimedia Commons

On August 31, 1955, a proposal for a research project on artificial intelligence at Dartmouth College was signed by John McCarthy (1927–2011),
Marvin Minsky (1927–2016), Nathaniel Rochester (1919–2001), and Claude
Shannon (1916–2001). At the time, the four scientists were working, respectively, at Dartmouth College, Harvard, IBM, and Bell Labs—one more
example of how productive collaboration among universities and companies can be.

The proposal stated:

> *We propose that a 2-month, 10-man study of artificial intel-
> ligence be carried out during the summer of 1956 at Dartmouth
> College in Hanover, New Hampshire. The study is to proceed on the
> basis of the conjecture that every aspect of learning or any other fea-
> ture of intelligence can in principle be so precisely described that a
> machine can be made to simulate it. An attempt will be made to
> find how to make machines use language, form abstractions and con-
> cepts, solve kinds of problems now reserved for humans, and improve*

*themselves. We think that a significant advance can be made in one
or more of these problems if a carefully selected group of scientists
work on it together for a summer.*

While these pioneers underestimated the complexity of the problems
to be solved, and generated expectations that ended up not materializing
over the short term, they nevertheless formed the general outlines for the
research to come.

During the 1970s, in a world deeply impacted by serious economic problems,
and at the end of the 1980s, with generic-use equipment replacing machines specifically developed to execute tasks related to artificial intelligence, very few people
were paying attention to what was happening in the area. It is estimated that by
1987, virtually the entire industry of machines that had been created to support
the development of applications in LISP (which stands for *list programming*) was
decimated. This language became popular with the success of the so-called expert
systems—an artificial intelligence technique based on sets of rules of the type "if
A, then B"; it was pioneered by the XCON (eXpert CONfigurer) system, developed in 1978 at Carnegie Mellon University in Pittsburgh, Pennsylvania.

The so-called artificial intelligence winter had begun: a promising field that
took a while to make good on its promises had caused its investors (including the
government) to lose patience and withdraw the funds needed for financing
the research and development work. The situation gradually started changing
in the last decade of the twentieth century and the start of the twenty-first, with
the positive perception of the potential of so-called intelligent computational
techniques combined with advances in processing speed, storage capacity, and
analysis of large sets of data.

An important discussion is underway about the emergence of artificial
general intelligence. Unlike the intelligent systems currently in use, which are
designed to serve specific purposes such as playing chess, recognizing a face,
or translating a text, artificial general intelligence could learn virtually anything, including *how to learn*. This would create a spiral in the accumulation
of knowledge that would know no bounds, and whose emergence would cause
unprecedented and unforeseeable changes in society's structure. Scientists have
not arrived at a common understanding around when this could be developed:
Some think it could be within just a few decades, others think it won't be until
the next century, and others think it will never happen.

MACHINES THAT CAN LEARN

Artificial intelligence research can be subdivided in different ways: as a function of the techniques used (such as expert systems, artificial neural networks, or evolutionary computation) or of the problems addressed (e.g., computer vision, language processing, or predictive systems). Currently, one of the most commonly used artificial intelligence techniques for the development of new applications is known as machine learning. In basic terms, machine learning seeks to present algorithms with the largest possible volume of data, allowing systems to develop the capacity to autonomously draw conclusions. A simple way to describe the process is as follows: If we want to teach an image-recognition system to identify a key, we show it the largest number of keys possible for its training. Then, the structure itself learns to identify whether subsequent images presented are or are not keys—even if the system never saw these images during its training.

Recognizing an image used to be a task in which humans had a clear advantage over machines—until relatively recently. Initiatives such as the ImageNet project, formulated in 2006, have served to significantly reduce this difference. Led by Chinese American researcher Fei-Fei Li, a computer science professor at Stanford University who also served as director of the Stanford Artificial Intelligence Lab (SAIL), the ImageNet project consists of a database with nearly 15 million images that have been classified by humans.

This repository of information is the raw material used to train the computer vision algorithms and is available online free of charge. To boost development in the area of computer image recognition, the ImageNet Large Scale Visual Recognition Challenge (ILSVRC) was created in 2010 where systems developed by teams from around the world compete to correctly classify the images shown on their screens. The evolution of the results obtained over less than a decade is proof of the extraordinary advances made in the field of deep learning (currently one of the most-used techniques in artificial intelligence, and a key enabler of—you guessed it—deep tech). In 2011, an error rate of 25% was considered good; in 2017, of the 38 teams participating, no less than 29 obtained an error rate lower than 5%.

For decades, the development of computer programs was based on the equation "rules + data = results." In other words, the rules were entered beforehand, input data was processed, and results were produced. But the paradigm used by systems based on deep learning is substantially different and seeks to imitate the way humans learn: "data + results = rules." Typically implemented through artificial neural networks (structures that are able to extract the characteristics necessary for the creation of rules from the data, and to produce results), these systems are on the front lines of platforms for facial recognition, voice recognition, computer vision, diagnostic medicine, and more. Once a sufficiently large set of examples (data) is presented with its respective classifications (results), the system obtains an internal representation of the rules—and becomes able to extrapolate the results for data it has not seen before.

DOING THE RIGHT THING

Although systems based on deep learning are able to improve the accuracy of virtually any classification task, it is essential to remember that their accuracy is highly dependent on the quality and type of data used during the learning phase. This is one of the biggest risk factors for the use of this technology: If the training is not done carefully, the results can be dangerous. In a 2016 study, three researchers from Princeton University—Aylin Caliskan, Joanna Bryson, and Arvind Narayanan—used nearly a trillion English words as input data. The results indicated that "language itself contains historic biases, whether these are morally neutral as toward insects or flowers, problematic as toward race or gender, or even simply veridical, reflecting the distribution of gender with respect to careers or first names."

Also in 2016, the monthly magazine of the Association for Computing Machinery (the world's largest international learning society for computing, founded in 1947) published an article by Nicholas Diakopoulos (a PhD in computer science from the Georgia Institute of Technology) entitled *Accountability in Algorithmic Decision Making*. If so-called intelligent systems do continue their expansion into different areas of business, services, and governments, it will be critical that they not be contaminated by the biases that humans develop, whether consciously or subconsciously. It is likely that the ideal model will involve collaboration among machines and humans, with the latter likely to be responsible for making decisions on topics with nuances and complexities not yet fully understood by models and algorithms.

The perception of the significance of future changes in practically all industries is reflected in the increase in investments in startups from the sector: According to the firm CB Insights, this figure went from less than $2 billion in 2013 to more than $25 billion in 2019.

Tech companies like Google, Microsoft, Apple, Facebook, and Amazon already incorporate intelligent techniques into their products and are moving toward a future where virtually all of their business lines will have a built-in machine learning component. This can apply to all types of applications: automatic simultaneous interpreting during a call, recommendations for whatever we want (or will want) to purchase online, or correct voice recognition in interactions with our cell phones.

One of the big challenges for companies is to define the best way of using this set of new techniques, which will contain probabilistic aspects in their outputs. In other words, the algorithms estimate a solution to a given problem, with no guarantee that it is actually the best solution. Either the process is robust and reliable, as a function of the quality of implementation and of the techniques used, or the results will be harmful to the financial health of the company in question.

The number of acquisitions of machine learning startups has been growing, led by the large tech companies, and, more recently, with participation from other sectors such as automotive, electronics, and industry. One of these transactions occurred in November 2016: the acquisition of the Canadian company Bit Stew by General Electric, for around $150 million. Bit Stew developed a platform to integrate and analyze data obtained from connected industrial devices. We will discuss this integration of physical objects with the digital world (also known as the Internet of Things) in the next chapter.

5

THE INTERNET OF THINGS
AND SMART CITIES

THE LATIN PREFIX *INTER* CAN BE TRANSLATED AS SOMETHING that joins items, something that lies between these items; the English word *net* is short for *network*. In other words, the Internet is the technology that enables the connection of several different networks, assuming an intermediate—or integrative—position among them.

What we have come to know as the Internet (with a capital *I*) originated in the 1960s, in the United States. The government, in particular the Department of Defense, needed a decentralized network to prevent an attack on a central point from taking down the entire communication structure among the computers of that time. Initial research efforts culminated in the ARPANET (Advanced Research Projects Agency Network), a precursor of the current Internet, which in late 1969 connected four computers—one at the University of California in Los Angeles, another at the Stanford Research Institute, another at the University of California in Santa Barbara, and one at the University of Utah.

The evolution of communication protocols and interoperability enabled the network to expand, although it remained restricted to academic and military settings up until the early 1990s. Since then, the Internet—which connects several networks of computers spread out around the world—has become an integral and essential part of the lives of billions of people, forming the basic infrastructure of the modern world.

THE INTERNET AND THINGS

For quite a while, computers were the only components that made up the Internet—but theoretically (and simplifying things a bit), any device able to "speak" Internet Protocol, or IP, could be connected to this global communications network: All that was needed was to structure and organize the transmission of information into digital packets and correctly dispatch each one of them. This is precisely what the pioneers of voice over IP technology (or VoIP) did in the early 1990s, developing telephones that—rather than connecting to the traditional, existing telephony network (with a high associated cost for international calls)—connected to the same data network as computers.

Over the following two decades, the expansion of broadband networks and increasingly universal access to the Internet, combined with the development of low-cost wireless sensors and circuits, facilitated the creation of a new and massive market: the Internet of Things (IoT).

Almost any piece of equipment or component can be considered to have the potential to be connected to the Internet. These run the gamut from household and personal devices, like cars, motorcycles, fridges, cameras, washing machines, air conditioners, lights, and coffee makers to heavy machinery, like airplane engines, locomotives, and drilling rigs to devices embedded in living things (people, wild animals, livestock, crops, and forests). All of these items can now send information in real time to any part of the world for subsequent analysis using algorithms developed specially for this purpose and that are able to handle unimaginable volumes of data. The global information provider firm HIS Markit estimates that, in 2020, more than 30 billion units were connected to the Internet, and that this number could exceed 75 billion by 2025. It is a complex network that connects people and machines in an efficient manner, in all environments.

It is reasonable to imagine that over the next few decades the impact of the IoT will be comparable to that of the Internet itself, or that of mobile telephony. We are talking about the transformation, through the use of sensors, of practically any item from the physical world into a digital entity that is capable of transmitting data regarding location and operational status, being remotely updated and monitored, being integrated, and operating in a collaborative manner.

As we discussed in the previous chapter, Moore's Law—formulated in 1965 by Intel cofounder Gordon Moore, and which remains relatively valid thanks to advances in the design and manufacture of microprocessors—predicts that the processing power of integrated circuits will roughly double every two years. The miniaturization of components and a reduction in production costs have also allowed a significant drop in prices: In 2004, the average cost of a generic IoT sensor was $1.30; the cost was $0.60 in 2016 and should be below $0.40 in 2021.

Add to this two more pieces: the layer of real-time analysis of the massive amount of information produced by these sensors and the development of artificial intelligence techniques for learning, extrapolation, and autonomous behavior. Together, these developments are already modifying interactions among people, among people and machines, and among machines and other machines—yet another feature of the Deep Tech Revolution going on before our very eyes.

In International Data Corporation's (IDC) 2016 study of 4,500 executives representing a wide range of businesses in more than 25 countries, more than half stated that the IoT is critical for the success of their brands. The motivation to increase investment in this technology, according to them, is connected to productivity, speed to market, and automation.

General Electric estimated in late 2012 that in the next two decades global gross domestic product (GDP) could increase by up to $15 trillion with the gains in efficiency, productivity, and cost effectiveness that will come with the connection of industrial machines to the Internet. For comparison purposes, the United States' GDP was about $21.4 trillion in 2019.

The full-force arrival of the connected world via the IoT (or, as it is called by Cisco, *the Internet of Everything*) will affect traditional business segments connected to the industrial, power, and manufacturing sectors. Much like the case of insurance companies and connected cars discussed in Chapter 2, companies can now obtain information on how their products are used, anticipate potential problems with equipment, and charge their clients based on the pattern of use—similar to how software companies charge licensing fees for the use of their programs (in a paradigm known as SaaS—Software as a Service).

MACHINE TALK

But challenges remain—from the interoperability of equipment and communication protocols to matters related to security and privacy. When we think of data networks that carry our digital communications—emails, webpages, photos, videos, voice, chats—we think of an infrastructure that serves the people who are exchanging the information. But this network, whose cornerstone was laid in the 1960s, is no longer serving only humans. Telecom companies can expect huge growth from communication among machines, with no human intervention—otherwise known as machine-to-machine, or M2M, communication.

INTERNET PROTOCOL: VERSIONS 4 AND 6

One of the most common versions of the protocol that all items connected to the Internet need to understand—the so-called IP, or Internet Protocol—is version four (IPv4) which went live in 1983. It consists of addresses that enable the data circulating over the network to be duly transferred from the sender to the recipient. You may have already seen IPv4 (IP version 4) addresses in the decimal format: four numbers from one to three digits, separated by dots—for example, 52.0.14.116. Each of these four numbers is represented by eight bits (sets of zeroes and ones) such that each complete address has 32 bits. Since each of these 32 bits has only two possibilities—either zero or one—there are therefore 2^{32} distinct combinations of addresses, which is equivalent to approximately 4.3 billion.

Even with different solutions for optimizing the use of the addresses available, the upper limit supported by IPv4 will not be enough for an increasingly connected world. The number of users connected to the Internet is growing rapidly and, more importantly, the number of devices—computers, modems, laptops, tablets, and cell phones—is growing as well. In its November 2016 report on mobility, Ericsson estimated that the world already has nearly four billion smartphone users and predicted that this number would reach 6.8 billion in 2022.

The problem of the lack of IP addresses has been the subject of development efforts since the mid-2000s, yielding version 6 of the Internet Protocol, which already coexists with version 4. There are several differences between the two, and the most important is the number of addresses permitted by the new model: 2^{128}, or 340 trillion followed by 24 zeroes (340,282,366,920,938,000,000,000,000,000,000,000,000).

Another critical aspect under heavy discussion among different groups and consortia is the standardization of messages exchanged by machines in a secure and reliable way. Autonomous vehicles, industrial machines, household equipment, body sensors, and other items are all set to be able to communicate not just among themselves, but also with the algorithms that will act on the information received.

The materialization of a truly connected world with not only networked computers but also networked physical devices, all of which are able to receive and transmit data, will set the stage for one of the biggest changes in the business world of the last few decades. The Internet's infrastructure is being adapted to allow an extraordinarily large number of pieces of equipment to be connected and integrated. At the same time, initiatives around the world are seeking to standardize models of communication among machines, making for a less complex and more efficient environment.

The reduction in manufacturing cost for sensors (the components responsible for capturing and transmitting information) has made it so that currently a single piece of equipment can have different types of sensors installed in it: for example, those that measure proximity, light, noise, temperature, pressure, and humidity; a gyroscope (which determines an object's spatial orientation); and an accelerometer (which determines the force being applied to an object).

The sensor manufacturing market is just one more area with a positive outlook, where a consistent growth of at least 10% per year over the next few years is projected.

THE COMMON SENSE OF SENSORS

Managing company-owned assets using information obtained via sensors is one of the first broad-scope IoT projects being adopted by companies. Fleets of vehicles, merchandise, equipment, supply chains, and clients can now be monitored and measured in real time. For example, a company can monitor the temperature of a perishable item and send an alert to the driver transporting it in the event the temperature approaches a critical level. The alert can specify which package needs to be replaced or what other measures should be taken. If the vehicle is autonomous, it needs only to be driven to the nearest service center. If this is not feasible, the item in question may need to be discarded at the delivery location.

Sensors can also monitor and optimize energy consumption, whether in elevators, heating and cooling equipment, or lighting. With the ability to sense how many people are in a given space, they can also adjust the temperature accordingly to make more efficient use of energy. Sensors can also safely and preemptively track the maintenance schedule of any given piece of equipment—a component of a machine, a turbine, an engine, or an oil exploration rig. With the ability to monitor parameters such as wear, time of use, and fatigue of the material, sensors notify operators as soon as any type of action becomes necessary.

The IoT also brings to life an element of fiction depicted in the 2002 movie *Minority Report,* directed by Steven Spielberg and based on a 1956 short story by writer Philip K. Dick (1928–1982): the personalization of the advertising industry. With smart sensors embedded in products, it is possible to identify the consumer through their cell phone (in the movie this identification is done using their retina) and thus direct the most relevant ads and promotions to them.

The increases in efficiency and reduction of waste—smart thermostats that actively control the temperature, sensors that can monitor the types of food that are being thrown away, better sorting and recycling systems—that are made possible through the sensors, data, and algorithms in use around the world promise significant productivity gains, potentially boosting the economy and the efficiency of all of the components of the productivity chain, including production, packaging, and delivery. The IoT revolution will impact businesses, homes, and cities and has the potential to make major improvements to the world population's quality of life.

IDC's 2016 study cited the biggest challenges to the effective use of the IoT as security, privacy, implementation and maintenance costs, IT infrastructure, and specific skills for this new market (in which the capacity to process significant volumes of data is critical). These challenges would already be relevant if we considered only the IoT applications for the corporate world, but when also taking into account the inevitable widespread application of the technology in households and in cities, it becomes essential to handle these matters properly.

Smart cities—basically, an urban area that collects data using sensors and applies the insights obtained to better manage its services—are a priority for governments and companies around the world. The market potential for energy, infrastructure, security, transportation, construction, and health care could reach $1.5 trillion in 2025, according to consulting firm Frost & Sullivan.

Population growth in urban centers—whether owing to migration from rural areas or to increased life expectancy—as well as behavioral changes in society, including heightened environmental awareness and new ways of consuming and using goods, are systematically increasing demands for transformations. The nonprofit organization Population Reference Bureau estimates that by 2050 the percentage of the world's population living in cities will have surpassed the current 55%, growing to nearly 70%, with this figure reaching 75% in developed countries. Considering the projected average global population growth over the next 30 years, we will be experiencing a migration toward cities of nearly 1.5 billion people—that's 50 million people per year, roughly the population of Colombia or South Korea.

Unprecedented urbanization will require greater efficiency and robustness in practically all cities' processes and services related to the IoT and sensors, creating business opportunities in several different sectors, like transportation, utilities, safety, and environment. Large-scale sensor projects will make it necessary to collect, transmit, analyze, and act on information in real time. Typically, four types of players will have critical roles: equipment suppliers, telecommunications providers, integrators, and service managers.

Equipment suppliers are charged with outfitting the city with sensors that can perform the specific functions required for a given project. Such sensors can measure, for example, water leaks, energy consumption, traffic volume, waste toxicity, or noise levels. The information collected is transmitted over the telecommunications infrastructure, focusing on the communication between the equipment and the companies providing the services. Integrators, as the name implies, combine the different services over a homogenous and consistent platform, and the managers of the services monitor and ensure the desired result.

SMART(EST) CITIES AND COUNTRIES

Power and traffic are recurring themes in city planning. One part of the development of the so-called smart grid (simply put, an electrical grid that uses sensors to better measure, dispatch, and control energy) involves the installation of new meters and household energy storage systems that make it possible for energy to be stored and then used at peak times (when the rates paid by the consumer are higher). Alternative and clean energy sources could also be incorporated. Visible manifestations of a smart city might be the dynamic control of traffic lights tied to vehicular volume via sensors on streets and the optimization of routes for medical services.

The IESE Business School at the University of Navarra in Barcelona, Spain, has been publishing a ranking of the world's smartest cities since 2014. Their

EVOLUTION OF URBAN POPULATION

1960

2017

2050 *

POPULATION IN
URBAN AREAS

OVER 90%

70 TO 89%

50 TO 69%

30 TO 49%

NO INFORMATION

*ESTIMATE

Figure 5.1. Sources: ourworldindata.org/urbanization; United Nations World
Urbanization Prospects 2018; others

index, called the IESE CIMI (Cities in Motion), considers criteria such as human capital (measured, among other things, by the expenditure on leisure, number of museums, theaters, schools, and universities), the type of governance, mobility and transportation, urban planning, the environment, economics, social cohesion, international projection (measured, among other things, by number of passengers per airport, number of hotels, number of McDonald's restaurants, and number of conferences and meetings), and technology. In 2020, 174 cities in 80 countries were analyzed; the top five were London, New York, Paris, Tokyo, and Reykjavik. Other cities that stood out were Basel (social cohesion), Bern (governance), Hong Kong and Singapore (technology), New York (urban planning and mobility and transportation), and Reykjavik and Copenhagen (environment).

The initiatives implemented by these and other cities work directly on issues that are critical to improving the quality of life of the population—which continues its migration from the fields to the world's cities. In just over two hundred years—between 1800 and 2020—the world's population grew from 1 billion to 7.7 billion inhabitants. According to the United Nations Fund for Population Activities, there will be nearly 10 billion of us by 2050.

But it is in Estonia—a country with just 1.3 million inhabitants, sandwiched between the Baltic Sea, Latvia, and Russia—that the idea of a truly digital nation is coming to life. After the collapse of the Soviet Union in 1991, Estonia chose to take a path that governments around the world are still striving for: digital management of services for its citizens. In 2000, the country designated the Internet as a basic human right, becoming the first to do so. Nearly 10 years later, countries such as Costa Rica, Finland, Greece, Spain, and France had done something similar, and in 2019 it was India's turn. In a 2011 report and then in a 2016 resolution, the United Nations reaffirmed that "the same rights people have offline must also be protected online," making an addition to Article 19 of the Universal Declaration of Human Rights.

In Estonia, which some call "E-stonia" because of its broad-reaching digital programs, children receive a digital ID card when they are born that serves as a sort of login for paying taxes, opening a bank account, and voting, for example. In just 20 minutes, an entrepreneur can open a business, without needing to leave home—and the country's ecosystem of innovation is still reaping the rewards of the sale of Skype (whose code was developed by three Estonians and acquired by Microsoft in 2011 for $8.5 billion). To stimulate economic growth, an e-residence service was created in 2014, whereby anyone can open and operate a business in Estonia, regardless of where they live.

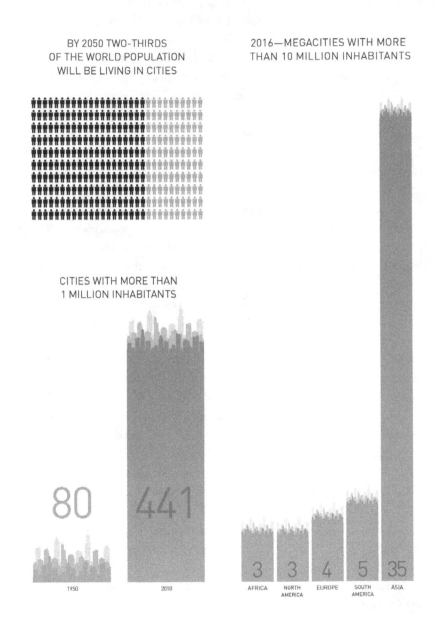

Figure 5.2. The megacity phenomenon. Sources: United Nations, Wikipedia

A politically motivated Russian cyberattack in 2007 had a profound impact on the country, and the Estonian government took measures to improve the robustness of its processes, including the opening of a data embassy in Luxembourg (where a secure copy of the government's information is kept) and the use of blockchain technology (a distributed and more secure database; more on that in Chapter 12).

A few cities now stand out on the world stage for their implementation of technological solutions for bringing intelligence to urban management. Singapore (actually a city-state) launched a program in 2014 called Smart Nation, whose initiatives include the installation of sensors and cameras around the city that feed into a platform called Virtual Singapore.

Despite the platform's benefits—which include the creation of jobs, business opportunities, and insights into the inner workings of the city and its population—there is a strong concern around information privacy and security. One could draw comparisons with Big Brother, a character created by Eric Blair (better known as George Orwell, 1903–1950) in his classic novel *Nineteen Eighty-Four*, in which the government blatantly monitors its inhabitants. In China, for example, it is estimated that by the end of 2017 there were already more than 170 million cameras installed on streets and in closed settings, along with a heavy investment in facial recognition systems, thus allowing the population to be located and monitored. And when it comes to security, as in any connected environment, there is a risk of digital attacks on the transportation system or the water, gas, or power supply systems.

Many cities provide access to the data collected (ideally respecting their inhabitants' anonymity) to assist in decision-making processes and engage the community. With more people tracking consumption, security, and pollution metrics (for example), faster decisions can be made. The information collected by a sensor's platform used in Barcelona, for example, is available to governments wishing to study it and apply it in their own smart cities. Information collected includes traffic, power, rainfall, air quality, noise levels, and more.

The fundamental difference between Singapore's smart city approach and those found in London, New York, or Barcelona is the high degree of centralization. Singapore marks the first time a government has been able to assemble such a complete set of information about its citizens, from the movement of vehicles to the use of cigarettes in areas where smoking is prohibited.

A MATTER OF SURVIVAL

With a growing world population and fewer inhabitants in rural areas, it is imperative for the world to significantly increase food productivity and cut down on farming waste, like water and air pollution due to chemicals or livestock production. According to an October 2009 report by the United Nations' Food and Agriculture Organization (FAO), global food production will need to increase by around 70% in the next two or three decades. By 2050, also according to the FAO, arable land area in developing countries will increase by 12%, but it will drop by around 8% in developed countries. This means that increased productivity will not be the result of the expansion of the rural frontier, but rather of improvements to processes connected to agricultural and ranching activities.

As stated by Czech-Canadian scientist and professor at the University of Manitoba, Vaclav Smil, if we were still working with the same productivity as at the start of the twentieth century, we would now need to use one-half of Earth's ice-free land surface for agriculture if it had not been for advances by two German chemists, who made it possible to produce fertilizers based on nitrogen, an essential element for the growth of virtually all crops.

AMMONIA: FRIEND AND FOE

Despite making up almost 80% of the air we breathe, the conversion of atmospheric nitrogen to ammonia—an essential element for plants and animals—proved to be a complex challenge. In 1909, German chemist Fritz Haber (1868–1934) developed a lab-based reaction (winning him the Nobel Prize in Chemistry in 1918) that combined nitrogen from the air with hydrogen derived from methane to produce ammonia. Together with BASF chemical engineer Carl Bosch (1874–1940), who would also go on to win a Nobel Prize in Chemistry in 1931 for the development of high-pressure chemistry, Haber developed the industrial-scale version of this process for agricultural application.

A sad note: During the First World War (1914–1918), Haber had worked on the development of chemical weapons for the German army. These activities supposedly led his wife, Clara Immerwahr (who earned the first

continued

PhD in chemistry from the University of Breslau—now the University of
Wrocław), to take her own life in 1915, just a few weeks before turning 45.

According to research company Euromonitor International, the sale of
industrialized foods—which prioritize ease of consumption—reached $2.4
trillion in 2014, and according to consulting firm Research and Markets, this
figure will top $3 trillion in the early 2020s. Considering that according to the
World Bank the global GDP in 2019 was about $90 trillion, and that agricul-
ture accounts for around 4% of this value, it is reasonable to estimate that the
food industry—including production and sales—represents approximately $7
trillion of economic activity.

Agriculture has a long history of advances and innovation with the consistent implementation of new processes and technologies. The mechanization
of the fields and the development of fertilizers, irrigation techniques, and
(more recently) genetic engineering are just a few examples of the forces that
have caused—in just the last 50 years—the productivity of different types
of crops to more than triple. According to data from the World Bank, US
grain production rose from 2,522 kg/ha (roughly 2,250 lb/ac) in 1961 to
8,281 in 2017.

The pursuit of efficiency in the fields has also come to ranching and the
production of meat. Ranching requires a significant amount of land (which then
cannot be used for agriculture), consumes a lot of water (an increasingly precious resource), and causes the emission of gases whose effects are as harmful
to the atmosphere as the burning of fossil fuels. To combat these problems, two
types of "meat" are either available or under development: plant-based meat and
cultured meat.

Plant-based meat is made from plant-based proteins, ideally offering
the consumer an experience close to that of real meat. While healthier for the
environment, this meat substitute—which has been gaining popularity thanks
to companies such as Beyond Meat (founded in 2009) and Impossible Foods
(founded in 2011 and whose mission is to "make our global food system truly
sustainable")—is not necessarily healthier for humans. It is *plant-based*, not
made from plants. The finished product contains zero cholesterol (it is mostly
soy, oils, and nutrient additives) but is a processed food and has more sodium
than its animal counterpart (a concern for blood pressure).

The second alternative method for producing meat is not yet ready for the general public but could potentially have a significant impact. Synthetic meat is created in a lab by cells harvested from the muscle of a live animal which are then cultivated in bioreactors in a highly nutritional growth medium (typically a cocktail that may contain elements like sugars, amino acids, and animal blood) to stimulate cell proliferation. This process is applicable to any type of animal from beef to fish. What's more, the bioreactors now being used by companies could eventually be placed in restaurants or even homes, dramatically modifying our present food distribution process. There are still significant technical challenges to overcome, but startups such as Just, Finless Foods, SuperMeat, Mosa Meat, and Memphis Meats (founded in 2015 by Marcus Post, a professor of vascular physiology and tissue engineering at Maastricht University in the Netherlands and a pioneer in synthetic meat production) are working toward this alternative.

Success in these advances will prevent a global food shortage, and the IoT is playing an important role in this field. Precision agriculture uses sensors that precisely map the topology of the land and collect information on soil, temperature, and humidity. Livestock can be monitored in detail via sensors, providing information on everything from the feed they consume to the health status of each animal. Images captured by drones or satellites (which are no longer restricted to government use) provide relevant information to the farmer, who can then integrate data into their decision-making process. Even tractors, which can now be autonomous, can collect data on the productivity of each acre, thus increasing efficiency and reducing waste in the field.

The Gartner Group estimates that more than half of the new processes and systems in the business world today include some IoT element. According to Machina Research, acquired by Gartner in 2016, the number of IoT devices in the rural segment will reach 225 million units by 2024 (versus 13 million in 2014). And evidence shows that the more connected a farm is, the greater its productivity and efficiency and the greater the reduction in its expenses related to, for example, power consumption and irrigation.

INSIDE OUR HOMES AND BODIES

But it is not only industries, cities, and farms that are implementing more connected and efficient environments. The trend toward monitoring and integrating many aspects of our daily lives via digital equipment is here. The structure of the

IoT includes the installation of sensors on items that need to be connected and the installation of sensors that enable objects from the physical world to be integrated into the digital world, generating an extraordinary volume of data and, once again, paving the way for new business possibilities.

The number of smart homes that are outfitted with appliances integrated with their manufacturers' networks, along with cameras, thermostats, doorbells, light switches, air quality monitors, and other devices that are accessible from anywhere, continues to grow. As we have seen, cities can manage traffic, pollution, and security, as well as improvements to infrastructure for power, lighting, water, and sewerage through monitoring and data processing. In a few places, garbage collection is already integrated between the public utility and the garbage cans themselves, whose sensors emit signals when they need to be emptied.

Our bodies are being integrated into the IoT structure via the wearables industry market; according to the firm Grand View Research, this market sector reached more than $32 billion in 2019 and is projected to expand at a compound annual growth rate of almost 16% until at least 2027. Using sensors in fitness wristbands, smartwatches, VR devices, and augmented reality (AR) devices, we can monitor and store data on body temperature, heart rate, calorie consumption, and sleep patterns. Sensors are likely to be incorporated into our clothing, or even into our bodies, over the coming years.

Personal sensors can make concrete contributions toward health. People with diabetes, for example, can monitor their blood sugar levels wearing a device that analyzes the fluids just under the skin using needles less than 0.5 mm (0.02 in) thick. Athletes can monitor themselves by wearable adhesives that analyze their sweat to track the risk of dehydration or cramping. The identification of patterns such as better nights after days with intense physical activity, or worse nights after eating a certain type of food, provides information that could lead to healthy behavior changes.

The constant monitoring of different metrics, such as heart and respiratory rate, blood oxygen levels, and temperature, provides benefits to both patients and doctors. The sensors (some of which are now biodegradable) allow patients to receive medical care not only in hospitals or clinics, but virtually anywhere. With certain types of sensors, medication may even be administered remotely with the release of substances directly into the blood stream. We can use technology not only to diagnose and treat illnesses, but also to anticipate and monitor events occurring in our bodies.

SENSORS AND DISEASE

According to data from the WHO (World Health Organization), member countries of the OECD spent 8.9% of their GDP on health care in 2018. Among the so-called BRICS—the group of emerging countries recognized for their economic potential—Brazil also reported 8.9%, followed by South Africa (8.2%), Russia and China (5.3% each), and India (3.9%). In the United States alone, health-care expenses represented 17.2% of GDP, or more than $3 trillion.

As stated by the WHO, in 2016, around 54% of the more than 56 million deaths in the world were caused by only 10 types of illnesses, and the leading causes of death since 2000 have been heart problems and strokes. Illnesses eradicated in the past few decades—such as smallpox and rinderpest—and diseases with significantly lower mortality rates today like malaria, mumps, and measles—are being replaced with an increase in cases of cancer, stroke, and degenerative diseases (such as Parkinson's and Alzheimer's). The ability to continuously monitor the symptoms and effects of diseases could fundamentally change the current state of affairs, as contact tracing has shown during the 2020 COVID-19 pandemic.

A paper published in January 2017 by the Stanford University School of Medicine provided a concrete example of the possibilities for the future of diagnostics. A group of 43 volunteers was monitored, and their individual data was analyzed and correlated. The scientists were able to precisely detect emerging signs of Lyme disease (which causes joint swelling and pain) and of inflammatory processes; they were also able to clearly distinguish the physiological differences between people who are sensitive or resistant to insulin. Considering that the wearables used in this study were general-purpose commercial devices—measuring things such as heart rate, blood oxygen levels, skin temperature, sleep, and calorie consumption—it is possible to imagine the advances that will be achieved using devices developed specifically for medical applications.

In spite of advances in wearable sensors, the health-care professional career is among those that are least vulnerable to automation. In the 2006 report *Working Together for Health*, the WHO indicated a global shortfall of over four million professionals, especially in the poorest regions of the world. This assessment, combined with increased life expectancy, only reinforces the importance of the evolution of health-related technologies, such as the ability to efficiently and autonomously diagnose health issues.

In the 1950s, global life expectancy was less than 50 years; in 2019 it reached about 73—and several countries have a life expectancy of more than 80 years (including Canada, Japan, Australia, and Finland). Scientific advances, public policies, educational and vaccination campaigns, and healthier habits are factors that have contributed to these changes. But the number of hospitals, especially in developed countries, is dropping. The United States had around 1.5 million hospital beds in 1975; in 2014, it had 900,000. The OECD reports the number of hospital beds in Europe dropped 1.9% every year from 2000 to 2010.

The average duration of each hospitalization has also fallen because of new medicines and less invasive procedures, and the focus of future health-care scenarios is to prevent and avoid hospitalization at all costs. It is no coincidence certain segments of health care, such as in-home care, are undergoing significant growth. In this scenario, the family, the person receiving care, and the health-care professionals share information using IoT devices such as smartphones, tablets, cameras, and sensors—and even devices into which the full dose of monthly medications can be preloaded and that can send alerts and messages to the patient, the doctor, the caregiver, or family members, according to the settings entered.

The integration of computer science, artificial intelligence, telecommunications, biology, and health services is permanently modifying our models in matters such as disease prevention, diagnosis, and treatment. MedTech (an abbreviation for "medical technology") encompasses the development of devices, therapies, diagnoses, medications, processes, and systems to care for the population's health and improve our quality of life.

This is an area with a heavy flow of investment into startups. According to data from CB Insights, in 2019 alone approximately $54 billion was invested in health-care startups all over the world. Biotechnology, one of the most promising and exciting branches in the broad landscape of life sciences, is the subject of Chapter 6.

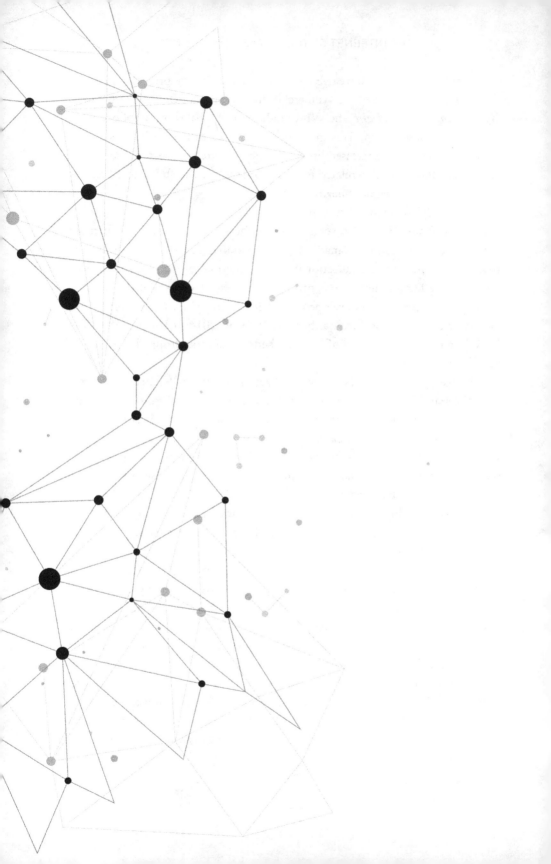

6

BIOTECHNOLOGY

WITH INCREASED LIFE EXPECTANCIES AND SIGNIFICANT reductions in the number of deaths due to contagious diseases, medicine's primary focus has turned toward chronic illnesses and those caused by the aging process. The marriage between technology and health is inevitable here, not only because of the costs associated with hospitalization and treatments, but also because a core strategy of medical practice in the twenty-first century is prevention.

Many entrepreneurs focused on innovation are engaged in creating businesses, products, and services for the health-care sector due in great measure to the size of the market and the applicability, importance, and economic feasibility of certain types of lab equipment. According to the Institute for Health Metrics and Evaluation, an independent population health research center at the University of Washington, global expenditures in health care went from $780 billion in 1997 to $7.9 trillion in 2017—a 10-fold increase in 10 years. This sum is equivalent to the combined 2019 GDPs of Germany, the United Kingdom, and Mexico. We've discussed wearables, which enable the monitoring of several different indicators, but those are only part of the picture.

YOUR HEALTH

Today's patients are very different from those of just a few years ago: With exposure to an unlimited amount of information, images, and videos, data that used to only be accessible in specialized publications is now a topic of discussion with our doctors. Several companies are developing ways to organize, provide access to, and optimize the flow of data among patients, health-care professionals, labs, and clinics. In addition to the need for the standardization of data, one of the most important questions is around information security: Our medical records are among the most personal and sensitive types of data.

Individualization is one of the key words for the future of medicine. Genetic testing by the company 23andMe (the number 23 in the company's name refers to the 23 pairs of chromosomes in human cells) can detect whether a patient is predisposed to certain types of diseases over the course of their lifetime. This knowledge allows for a more robust mapping of an individual's profile and directed preventive exams, but it is very important to note that the results of these genetic tests are only for reference; they are not definitive clinical diagnoses.

Advances in biochemistry also increase the accuracy and the volume of information that can be obtained from blood samples, while the use of artificial intelligence techniques in imaging boosts the speed and accuracy of the diagnosis of any anomalies detected. In a 2016 article written by researchers from Google and several universities, published in the *Journal of the American Medical Association,* exciting results were presented on a new technique for detecting signs of diabetic retinopathy. This illness can cause blindness if not treated, and the number of doctors who are sufficiently specialized in retina image analysis is limited. The research team developed a system, based on machine learning, that was taught to recognize early signs of the illness and thus signal the need for treatment—and its accuracy rate, tested on thousands of cases, was in line with the opinions of the specialists who were consulted. Even when medicine does not manage to halt a condition that can lead to a disability—whether due to genetic causes or to accidents, whether physical or mental—technology is available to assist and improve the quality of life of the affected person.

Of all the technological advances the world will continue to experience, few surpass—in terms of benefits to humanity—the advances related to life sciences, such as biotechnology, epidemiology, and medicine. The industries connected to these fields, like pharmaceuticals and medical equipment, will continue undergoing significant changes over the coming decades, thanks to innovations like those that make it possible to simulate the effects of new drugs on the body, accelerate validation studies for chemical compounds, and edit the genetic code of a living being.

EDITING DESTINY

The turning point in the history of modern genetics—the science that studies genes, their behavior, and the mechanisms of heredity and whose name comes from the Greek word *genesis*, which means origin—took place when Augustinian friar Gregor Johann Mendel (1822–1884) started working at a 2-ha (5-ac) garden at St. Thomas's Abbey in the current Czech Republic. Mendel crossbred peas and carried out experiments on nearly 30,000 plants between 1856 and 1863, which led him to the discovery of the laws of inheritance. Journalist and writer Robin Henig (author of *The Monk in the Garden*) reveals that Mendel had to modify his original research plan because the bishop did not want anyone studying animal reproduction in the abbey, and Mendel's initial subjects had been rats.

The laws of Mendelian inheritance were revolutionary in their understanding of how one generation's characteristics are passed on to the next. But despite the groundbreaking nature of his discovery, it took nearly 35 years for it to gain the recognition it deserved in the scientific community. Structures in cells, called chromosomes (from the Greek *chromosoma*, literally "colored body," because of the color these structures take on when dyed), were visible under a microscope during the cell division process. At the start of the twentieth century, a few researchers connected these structures to Mendel's work and concluded, correctly, that chromosomes are responsible for an organism's genetic predisposition. German biologist Theodor Boveri (1862–1915) and American geneticist Walter Sutton (1877–1916) independently reached the same conclusion, which led to the Boveri-Sutton chromosome theory of inheritance.

In 1944, scientists Oswald Avery Jr. (1877–1955), Colin MacLeod (1909–1972), and Maclyn McCarty (1911–2005) carried out an experiment that determined that DNA—the structure from which chromosomes are made—is responsible for an organism's genetic load. DNA (an abbreviation for deoxyribonucleic acid) was first observed in 1869 by Swiss doctor Friedrich Miescher (1844–1895), and its now universally recognized double-helix shape was discovered almost 85 years later in a sequence of events still surrounded by controversy.

In May 1952, Raymond Gosling (1926–2015) was a PhD student of Rosalind Franklin (1920–1958) at King's College in London. Under her supervision, Gosling obtained an image of DNA using an x-ray diffraction technique. A short time later, with Rosalind Franklin's departure from the university, molecular biologist Maurice Wilkins (1916–2004) became Gosling's supervisor, and under instructions from the director of the research group, physicist John Randall (1905–1984), he shared the image (without Franklin's permission) with his colleague, molecular biologist James Watson. At the time, Watson was working with Francis Crick (1916–2004) at the University of Cambridge, in England. Together, using the data obtained from the photo, they published the correct model of the DNA molecule in a 1953 issue of *Nature* magazine, paving the way for the cracking of the genetic code.

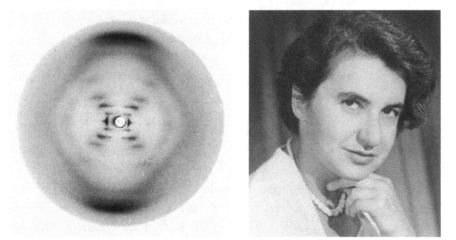

Figures 6.1 and 6.2. "Photo 51," the image obtained by Raymond Gosling (1926–2015) during his time as a PhD student of Rosalind Franklin; Rosalind Franklin (1920–1958). Sources: King's College London Archives and Jewish Chronicle Archive/Heritage-Images

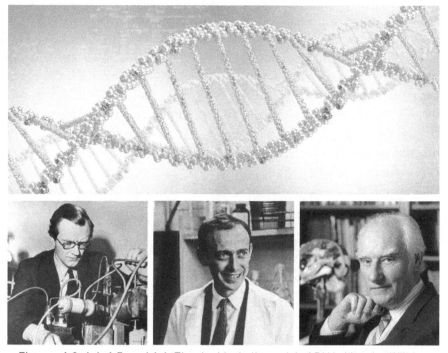

Figures 6.3, 6.4, 6.5, and 6.6. The double-helix model of DNA. Maurice Wilkins (1916–2004), James Watson, and Francis Crick (1916–2004), winners of the Nobel Prize in Physiology or Medicine 1962. Due to prize regulations stating that the award cannot be granted posthumously, Rosalind Franklin was not included. Sources: Shutterstock, Wikimedia Commons, Courtesy of the National Library of Medicine, Siegel RM, Callaway EM, Francis Crick's Legacy for Neuroscience— PLoS Biol (photo by Marc Lieberman)

British naturalist Charles Darwin (1809–1882) expanded on the ancient Greek theory of pangenesis to try and explain the transmission of characteristics from one generation to the next. The pangenesis theory purported that animals emit tiny particles over the course of their lives, which are incorporated into their descendants in the reproductive process. Historians and scientists still debate what impact Mendel's discoveries would have had on science had Darwin been familiar with them when he wrote his theory on the origin of species and on natural selection. In fact, the combination of both discoveries ended up leading to the field of evolutionary biology, and one of the first scientists to combine heredity with natural selection was English statistician Ronald Fischer (1890–1962). Fischer had as a primary mentor Leonard Darwin (1850–1943)—one of Charles Darwin's 10 children.

In 1909, Danish botanist Wilhelm Johannsen (1857–1927) published the term *gene* for the first time, in opposition to the concept of the *pangene*. Eleven years later, the word *genome* was published for the first time by German botanist Hans Winkler (1877–1945), when he combined *gene* and *chromosome*. Once the general mechanisms of heredity were reasonably well understood, the search began for a detailed understanding of the process—in other words, how each gene can influence the development of an organism through its specific function. To gain this understanding, it was necessary to obtain the complete sequence of all genes of the organism being studied, particularly nucleotides—the molecules that make up DNA.

THE HUMAN GENOME PROJECT

The most important part of nucleotides in genetic terms are the triphos-phates: adenine (A), thymine (T), cytosine (C), and guanine (G). In 1972, Belgian molecular biologist Walter Fiers (1931–2019) sequenced a gene, and in 1976 he sequenced a virus. Less than 30 years later, in 2003, the Human Genome Project, financed by the government of the United States between 1990 and 2003 for a total inflation-corrected cost of around $5.5 billion, completed the sequencing of the 23 pairs of chromosomes of a human being, analyzing nearly twenty thousand genes and mapping over three billion base pairs (combinations of A, T, C, or G). The development of new technologies has caused the cost of sequencing a human genome to fall dramatically, and databases containing the complete genomes of different types of plants and animals are available, opening up a broad field of research that will provide decades of research work (please refer to Figure 6.9.).

Recently, the foundation for one of the biggest revolutions in the history of genetics was laid, thanks to studies of microbes and bacteria whose genetic code had a DNA sequence that repeated several times. These repetitions were inter-rupted by sections of DNA that contained the genetic material of viruses that, at some point during the life of the organism, had tried to attack the original cell.

These sections ended up serving as a memory for the cell's defense system because they stored the identity of the cell's aggressor; this configuration of interrupted repetitions was named *clustered regularly interspaced short palindromic repeats*, or CRISPR (pronounced "crisper"). Combined with a set of enzymes known as *Cas*, it is possible to use the genetic material present in the spaces between the repetitions—that is, the genetic material of the invading viruses—to guide the enzymes to cut out the virus and neutralize it.

Figure 6.7. Walter Fiers (1931–2019), Belgian molecular biologist and pioneer in the area of genetic sequencing, the price for which fell even faster than expected under Moore's Law.
Source: VIB.be

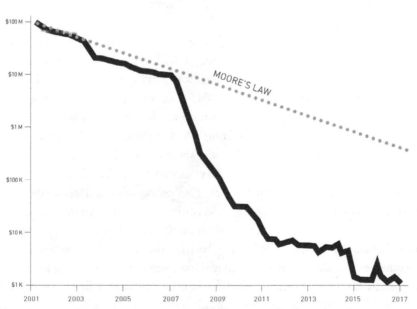

COST PER GENOME

Figure 6.8. Source: National Human Genome Research Institute

The tremendous revolution lies in the capacity the scientists acquired, using this method, to cut out any DNA sequence in the exact location desired—and replace the removed sequence with another, predefined one. In other words, complex illnesses, caused by different genomic alterations, could be studied in a much more efficient way, and eventually a series of genetic problems could be corrected before causing harm to the organism. Thanks to this new technique—for which French professor Emmanuelle Charpentier and American biochemist Jennifer Doudna were awarded the 2020 Nobel Prize in Chemistry—procedures that were once unviable would now be at the disposal of science.

According to a report published in March 2017 by Grand View Research, the market for genome editing is projected to exceed $8 billion by 2025—the technology has allowed scientists to decode, in record time, the genome of the SARS-CoV-2 (COVID-19) virus, responsible for the 2020 pandemic. Just by looking at Figure 6.8 it is pretty clear that something massive is happening: It is the perfect embodiment of the Deep Tech Revolution we have been talking about. Multiple technologies have converged and created an accessible, revolutionary, and flexible business case based on decades of research. Between 2018 and 2019 alone, investors have poured more than $30 billion into biotech startups—and it does not seem that this trend will end anytime soon.

But, of course, such powerful technology brings important moral questions. Eric S. Lander, professor at both Harvard and MIT schools of medicine, alerted the scientific community to the ethical risks of writing new genetic codes (especially in human embryos) in the July 2015 issue of the *New England Journal of Medicine*, but he also underlined its benefits once the technology is fully mastered. Hemophilia may be addressed by editing blood stem cells, and the deactivation of a certain gene in retina cells could prevent some types of blindness.

Another example relates to HIV, the virus that causes AIDS (acquired immunodeficiency syndrome), which does not affect people who do not have the CCR5 gene. In theory, removing this gene could confer immunity to an organism. At the end of 2018, it came to light that a team from the Southern University of Science and Technology in Shenzhen, China, was recruiting couples that would allow their babies' genetic code to be edited making them not only resistant to HIV, but also to smallpox and cholera. There is strong evidence that the first Chinese babies who underwent this procedure have already been born—sparking strong reactions among the medical community and the general public due to ethical concerns.

GENOME SIZE FOR SELECT ORGANISMS

SPECIES	GENOME SIZE
GENLISEA TUBEROSA (carnivorous plant)	60 MILLION BASE PAIRS
DROSOPHILA MELANOGASTER (fruit fly)	175 MILLION BASE PAIRS
TETRAODON NIGROVIRIDIS (green spotted puffer)	342 MILLION BASE PAIRS
OPEROPHTERA BRUMATA (moth)	638 MILLION BASE PAIRS
LEPIDOTHRIX CORONATA (blue-crowned manakin)	1 BILLION BASE PAIRS
BOS TAURUS (cow)	2.7 BILLION BASE PAIRS
HOMO SAPIENS (human being)	3.3 BILLION BASE PAIRS
PERIPLANETA AMERICANA (cockroach)	3.4 BILLION BASE PAIRS
RANA CATESBEIANA (bullfrog)	6.2 BILLION BASE PAIRS
AMBYSTOMA MEXICANUM (salamander)	32.4 BILLION BASE PAIRS
PROTOPTERUS AETHIOPICUS (marbled lungfish)	133 BILLION BASE PAIRS

Figure 6.9. Source: National Center for Biotechnology Information

SYNTHETIC BIOLOGY

Technologies that enable us to edit and synthesize DNA have given rise to the area of synthetic biology, which allows us to effectively create, modify, and program organisms to act according to our needs. The winners of the Nobel Prize in Physiology or Medicine in 1978 (Swiss geneticist Werner Arber, American microbiologist Daniel Nathans [1928–1999], and American microbiologist Hamilton Smith) worked on restriction enzymes that are able to cleave the target DNA in specific locations (not unlike CRISPR). This area, according to the SynBioBeta network, saw nearly $4 billion in new investments during 2018.

Advances in the artificial synthesis of genes (demonstrated in 1972 by a team led by biochemist Har Gobind Khorana [1922–2011]) and synthetic biological circuits, which, like their electronic counterparts, can be designed to execute predefined logical functions, laid the foundations for a new frontier that has already started affecting food (with the development of synthetic meat), fertilizers (with the use of microbes specifically created to provide the nitrogen available in the soil to plants), and in new materials like bioplastics (that may replace petroleum-based plastics).

WHO WANTS TO LIVE FOREVER?

A dramatic global change in population pyramids is taking place all around us in a phenomenon called the Silver Tsunami. According to the US National Institutes of Health, in 2050 nearly 16% of the world's population (something like 1.5 billion people) will be 65 or older, compared with only 8% (525 million people) in 2010.

The growth in the 65+ age group is set to occur quite differently depending on the geographic region: In developed countries it will be around 70%, whereas in developing countries this number will top 250%. This has significant economic implications for government and for society, not only due to the number of people affected, but also because of the growing costs of health care. Irrespective of country, social class, or sex, we are living longer. In 2015, for example, the country with the highest life expectancy (according to data from the WHO) was Japan, at nearly 84 years, with children under 14 accounting for just 13% of the population,

POPULATION PYRAMIDS: YESTERDAY, TODAY, AND TOMORROW

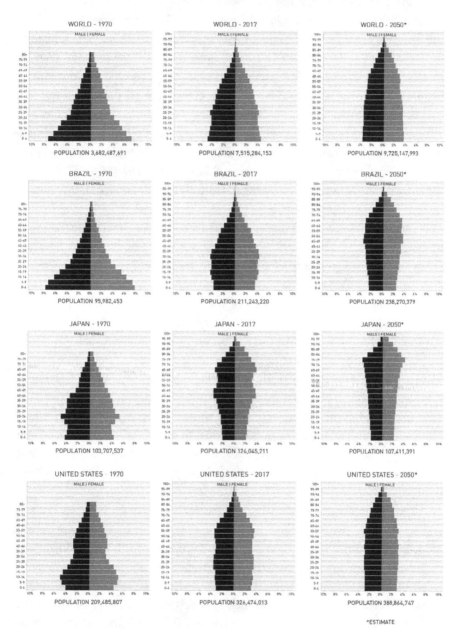

Figure 6.10. Source: PopulationPyramid.net

versus 26% for adults over 65. By 2011, according to Japan's largest diaper manufacturer, sales for adults had already exceeded sales for babies.

In a *Science* magazine article published in 2002, authors Jim Oeppen (from
the University of Cambridge) and James Vaupel (from the Max Planck Institute
for Demographic Research, in Germany) noted that over the past 150 years, we
have been increasing our life expectancy at a rate of approximately three months
per year. According to the WHO, life expectancy around the world increased
by 5.5 years between 2000 and 2016. Even more impressive is the increase in
a little-discussed number, known as life span. Until 1960, biologists' estimates
were that the life span of a human—that is, the maximum age that could be
reached—was 89 years. Fifty years later, this value was revised to 97, and with
the increase in the number of centenarians throughout the world, this number
is set to be adjusted again soon.

Human life expectancy is increasing thanks to several factors, and, according
to economist Johannes Koettl of the Brookings Institute, the most important
factors are regenerative medicine and organ transplantation. Antibiotics, vaccines, and the progress achieved thanks to hygiene measures have improved
our quality of life, and replacement or regeneration of vital organs has had a
dramatic impact on the life span of the beneficiaries.

In 2012, Steven Cave (PhD in Philosophy, University of Cambridge) published *Immortality: The Quest to Live Forever and How It Drives Civilization*.
Cave argues that every civilization can be characterized by different technologies that increased human life span: agriculture, to ensure food during the entire
year; clothing, to protect people from the cold; engineering, to provide shelter;
arms, for hunting and defense; and medicine, to treat injuries and diseases. His
reasoning leads to the conclusion that it is not solely the pharmaceutical, hospital, and medical equipment industries whose primary purpose is to preserve
our health.

Over two millennia, each civilization has set standards for clothing, housing,
and implements, while seeking to prolong the lives of its citizens. Scientific
progress has enabled medicine to achieve extraordinary feats, thus reducing suf-
fering and improving people's lives. The current state of medical science is the
result of centuries of technological advances, including the creation of the first
hospitals in the fourth century, the understanding of human anatomy, the development of vaccines and anesthesia, the construction of equipment to see inside
the human body, the understanding of the mechanisms of degenerative disease,
organ transplantation, and the mapping of the genetic code. These are our weapons in the fight against our bodies' enemies.

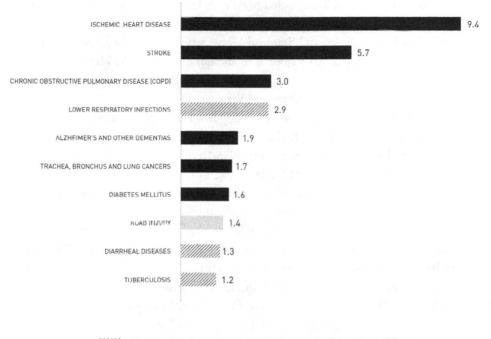

NEW ENEMIES

GLOBAL RANKING
CAUSES OF DEATH IN 2016
(MILLIONS)

ISCHEMIC HEART DISEASE	9.4
STROKE	5.7
CHRONIC OBSTRUCTIVE PULMONARY DISEASE (COPD)	3.0
LOWER RESPIRATORY INFECTIONS	2.9
ALZHEIMER'S AND OTHER DEMENTIAS	1.9
TRACHEA, BRONCHUS AND LUNG CANCERS	1.7
DIABETES MELLITUS	1.6
ROAD INJURY	1.4
DIARRHEAL DISEASES	1.3
TUBERCULOSIS	1.2

///// COMMUNICABLE, MATERNAL, PERINATAL, AND NUTRITIONAL CONDITIONS

▬▬▬ NONCOMMUNICABLE DISEASES

▬▬▬ INJURY

Figure 6.11. Source: *Global Health Estimates 2015: Deaths by Cause, Age, Sex, by Country and by Region*, 2000–2015, World Health Organization (Geneva, 2016)

Never has so much research been done on how we age and how we can try to delay this process. The organism's decay process has come to be seen as a disease itself. Rising life expectancies move trillions of dollars in research, diagnosis, medical procedures, equipment, and medicine, and longevity has taken up a prominent position in research centers, universities, and privately held companies around the world. The Buck Institute, founded in 1999 as the first private research institute focusing exclusively on aging; the Mayo Clinic, founded in 1889; and Calico (California Life Company), founded by Google in 2013, come to mind.

Promising research is connected to what is known as cellular senescence. In the early 1960s, Leonard Hayflick (of the medical schools of the University of California San Francisco and Stanford University) and Paul Moorhead (of the University of Pennsylvania's school of medicine) discovered the Hayflick Limit, which is the number of times a normal human cell can divide and, therefore, renew itself. The number is somewhere between 40 and 60.

In very simplified terms, when our body has a dysfunctional cell, the neighboring cells transmit a signal that allows our body to correct the issue or, in the case of cells that are harmful to the organism, remove the cell. When senescent cells send a signal that something is wrong too frequently, it causes constant inflammation that, in turn, causes several illnesses associated with degeneration and the passage of time (for example, Alzheimer's, arthritis, and heart disease). Scientists believe that if they can manage to prevent these signals from being sent by senescent cells, a series of chronic problems associated with aging will vanish. With investment into research on stem cells (which have the potential to develop into many different types of cells and serve as the body's repair system) it is possible to imagine the production of healthy tissue and organs, which could be made using 3D printers and transplanted into the patient, combined with noninvasive treatments based on nanotechnology (the manipulation of materials on an atomic or molecular scale).

3D PRINTING

WITH 3D PRINTING BECOMING PART OF OUR DAILY LIVES, IT may be surprising to know it has existed for several decades. Technical advances and lower prices have steadily made it available to a greater number of users and transformed the concept into an important pillar supporting the development of prototypes and products. The first patent applications were recorded in the 1980s for rapid prototyping technologies with the aim of building models of industrial equipment and components. Just 10 years later, the three most important techniques had been created and patented.

THREE 3D TECHNIQUES

In 1981, Hideo Kodama, of the Nagoya Municipal Industrial Research Institute, developed the first 3D printing methods using photopolymers (light-sensitive, repeating chains of molecules). However, the complete specifications for Kodama's patent were not submitted within the 12-month deadline after the initial application, so no patent was granted.

continued

In 1984, French innovators Alain Le Méhauté (of the Alcatel Alsthom Group), Olivier de Witte (from an Airbus subsidiary called CILAS and the Safran Group), and Jean Claude André (from the French National Center for Scientific Research) applied for a patent for the stereolithography (SLA) method. Stereolithography (or photosolidification) uses a beam of light that induces the linking of molecule chains, forming the polymers that will represent the complete 3D object. The researchers' employers did not envision commercial applications for the technology, and, as in Kodama's case, no additional commercial development followed.

But US engineer Chuck Hull's story was different. Just three weeks after the French application was submitted, he applied for a patent that was granted in 1986. That same year, Hull founded the 3D Systems Corporation that led to the commercialization of the first 3D printers. Fifteen years later, the company acquired a competitor, DTM, to handle the 3D printing technique called selective laser sintering (SLS). The patent for SLS was granted in 1989 to Carl Deckard and his academic supervisor, Joseph Beaman, both from the University of Texas at Austin. SLS uses a laser to group porous material in a specific order, creating a 3D object.

In 1989, mechanical engineer Scott Crump applied for a patent for fused deposition modeling in which the printer heats up and melts the material (typically metal or plastic) that is deposited in precise layers to create the desired object. That same year, Crump founded Stratasys, headquartered in the state of Minnesota.

In industries where the manufacture of a finished product using a mold requires the removal of excess material, the changes introduced by 3D printing are significant. Not only do physical molds become unnecessary, but raw material waste and energy consumption are both reduced. The technology also enables production in small batches, which would have been unfeasible in traditional assembly lines created to produce millions of exact copies of a given product. Typically, 3D printing is additive by nature: An object is produced via the deposition of very thin layers of material, one over the other, following a digital template in an operation controlled by a computer.

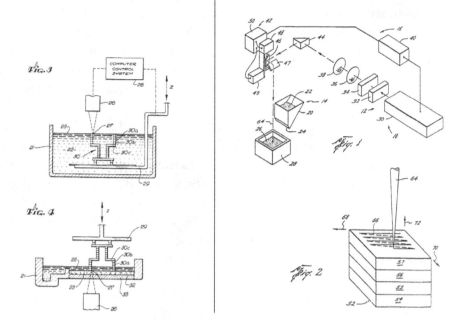

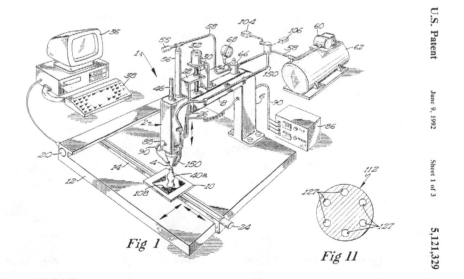

Figures 7.1, 7.2, and 7.3. The patents belonging to Hull (1986), Deckard (applied for in 1990), and Crump (1992), pioneers of 3D printing techniques.
Source: Google Patents

An exciting advantage of 3D printing is the power to customize products, from cups, clocks, door stops, models for hobbyists, and art reproductions to specialized parts for engines, wings, cars, or medical equipment—like the ventilators that were in short supply during the COVID-19 crisis. Providing specs for an article suddenly becomes simple and inexpensive since only the digital version needs to be modified. Once the virtual model is approved and the material selected, the file is sent for printing and the finished product may be delivered within days or even hours.

Technology's gradual improvement has enabled a reduction in lead time and a decrease in warehousing costs (with a digital inventory) in a wide range of sectors. Back in 2015, for example, Unilever's Italian division said they were able to cut lead times for prototype parts by 40%, and in 2018 Audi accelerated their design verification process using 3D printing as well. According to the results of the 2019 "State of 3D Printing" annual report published by Sculpteo, out of approximately 1,300 respondents surveyed about the top three benefits of 3D printing, 69% mentioned the handling of complex geometries, 47% mentioned quick iteration, and 42% mentioned lead time reduction.

In April 2019, Ernst & Young surveyed 900 companies in Europe, Asia, and North America about the global 3D printing market in sectors like aerospace, automotive, construction, consumer goods, electronics, life sciences, and transportation. According to the final report, 3D printing is quickly escalating: Two out of every three businesses claimed to already have experience with the technology (compared to just one in four in the previous report, published in 2016); in Asia, about 80% of the 130 participating companies have used 3D printing. Even more telling, 40% of all respondents have in-house systems in place.

It is not by chance, then, that several research companies estimate that the global expenditure in 3D printing equipment, software, and services will increase from around $10 billion in 2018 to more than $35 billion in 2024.

PRINT "ME"

Human organ transplantation using the patient's own cells offers stunning possibilities as it eliminates the risk of rejection and the need to wait for a matching donor. 3D bioprinters use a biodegradable mold to hold tissues and a gel saturated with cells that are permeated by a web of capillary vessels that receive nutrients and oxygen after implantation. Methods involving skin, cartilage, bladders, muscles, and urethras have already been tested, and initial results promise an alternative path for organ transplantation. Affordable 3D-printed arms and hands are now a reality thanks to the work of, for example, the e-Nable network of volunteers, as well as the development and printing of legs and feet using a digital template from a healthy member.

In 1982, US resident Hugh Herr had to have both legs amputated below the knee after spending three nights in temperatures below –25°C (–13°F) when he was caught in a snowstorm on a climbing trip. As a result, he decided to find a solution: He earned a master's degree in mechanical engineering from MIT, a PhD in biophysics from Harvard, and a postdoc in biomedical devices at MIT. One of his achievements includes an ankle-foot prosthesis that allows amputees to walk with a natural gait. His research, combined with the work of many others in the advancement of 3D printing, is creating a more natural and less uncomfortable integration between prosthetics and the human body.

PRINTING LUNCH

Global population is set to reach 10 billion inhabitants by 2050, and as we have seen in Chapter 5, we need to make sure we can sustainably feed a planet with that many people. Precision agriculture combines sensors, satellite images, software, and automation to increase productivity and efficiency in the fields. But will this be enough?

According to data from the United Nations FAO, the world food production index (which measures productivity) more than tripled between 1961 and 2013, while the percentage of land under cultivation remained about the same. Technology has enabled us to produce more food using the same amount of space while keeping prices affordable for most of the population—although shamefully, even in 2019, world hunger affected nearly 1 person in 10.

But food production itself causes significant environmental impacts. Researchers led by Professor Gidon Eshel from Bard College in New York attempted to quantify these impacts in 2014 in a study published by the US National Academy of Sciences. It showed that ranching accounts for around 20% of the total emissions of gases that cause the greenhouse effect (the name given to the rise in global temperatures caused by an increase in the concentration of pollutants in the atmosphere).

FOOD PRODUCTION
VS.
AGRICULTURAL LAND AREAS

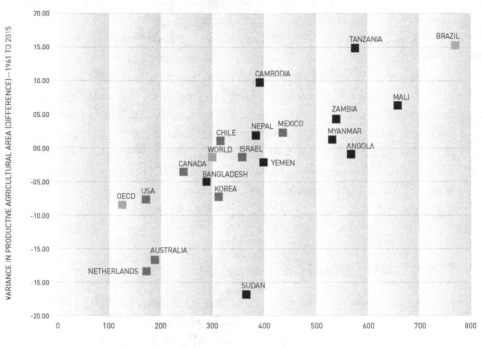

Figure 7.4. Sources: World Bank, United Nations,
Organization for Economic Cooperation and Development (OECD)

The study further stated that the production of beef when compared to other types of food production (dairy, chicken, pork, and eggs) requires 28 times more land and 11 times more irrigation, while emitting five times as much harmful gas into the atmosphere. This is why some environmentalists say they would rather see a vegetarian driving a heavily polluting car than a non-vegetarian who only uses a bike as their method of transportation.

In the same way that regular printers use ink cartridges, 3D printers can use plastic, sand, concrete, metal, organic materials, ceramics, or even food as raw materials to construct the product. And construct is a suitable word: Its origin is from the Latin—the prefix *con* means "together," and the verb *struere* means "to pile." 3D printing is the act of piling up the layers that are deposited during the creation process.

Different types of materials can be used as the "ink" to print food: extracts from organic foods, gelatin, protein concentrates (often algae-based), chocolate, salt, sugar, and flour are just a few examples, but in the future there are sure to be more possibilities. This prospect has many advantages: The cartridges can be transported efficiently and stored for longer. Countries with a food shortage—whether due to war, pollution, or natural disasters—could receive the raw materials and print out their meals.

In their 2013 book *Fabricated: The New World of 3D Printing*, Hod Lipson and Melba Kurman mention the potential connection between our personal food printer and the sensors that might monitor our bodies, thus enabling the "printing out" of a meal made of exactly the nutrients we need at any particular time and that would address allergies or other specific restrictions. Important technical questions must be answered over the next few years before consumers have their own food factories in their homes, but the journey toward this reality has begun.

THE MAKER MOVEMENT AND CREATIVITY

One of the most fascinating consequences of the economic accessibility of relatively sophisticated equipment such as 3D printers is that an entire generation—the maker generation—now has the means to design, develop, test, manufacture, and use their own creations. This generation has also benefited from the Internet culture of open-source architecture for software and hardware that makes it possible for the community at large to make improvements and extensions and correct possible problems found, for example, in open-source software—a type of software for which the source code is freely available. Source code is the set of instructions that defines how a particular computer program should behave. The Linux operating system (which has been evolving since 1991) and Android (evolving for more than 10 years) are famous examples of platforms whose respective source codes are partially open.

Synthetic biology uses the Synthetic Biology Open Language (SBOL); in the field of hardware we have the open-source hardware movement—analogous to what also exists for software but applied to the world of physical devices. This movement has already produced development and prototyping platforms such as the Arduino kit, a microcontroller (the brain of an electronic device) that can be manufactured and distributed freely and that supports the development of projects by makers, with a strong link to the IoT market. With a bit of technical knowledge, it is possible to build a prototype, whether for personal use or for testing the market potential of a given invention. Combine this with access to online education—technical courses, educational videos, classes from some of the world's best universities—and the efficacy of working collaboratively thanks to the Internet, and we have the foundations for the creation of a structured, growing movement.

The maker movement prioritizes creativity—one of the primordial human characteristics that artificial intelligence researchers seek to understand and reproduce in machines. As we have seen, labor market dynamics over the coming decades will favor jobs in which the capacity to improvise, create, and innovate are determining features. Until recently, the term *do-it-yourself* was associated with people who could stop a leak, paint a wall, or remodel a room or a house, but the concept has now become much broader: Makers are people who customize the world around them and, by modifying it, create their own reality.

Maker communities freely share their ideas and meet in both virtual and physical settings to share experiences, collaborate on projects, and present their ideas. According to *Popular Science*, in January 2016 there were around 1,400 makerspaces around the world—a 14-fold increase in 10 years (a makerspace is basically a place where users build, design, tweak, or invent existing or new objects or devices). Fab Foundation, created in 2009 as an outgrowth of an MIT program, seeks to connect a network of fablabs (fabrication labs) from dozens of countries.

The United Nations reports that between 1970 and 2010 the manufacturing sector's share in global GDP fell from approximately 26% to 16% due to automation and increased productivity, similar to what had happened earlier in the agricultural sector (which also saw a significant reduction in terms of number of jobs and contribution to GDP). However, due to advances in artificial intelligence techniques and greater processing power and storage capacity of computers, we are watching activities once done only by humans now being done by machines. This may be part of the reason why the maker movement has been gaining strength around the world, and that it is considered an important trend in the business world.

Writer and entrepreneur Chris Anderson published an article in *Wired* in 2010 called "In the Next Industrial Revolution, Atoms Are the New Bits." He argues that when the Internet was becoming popular, it was very easy to share a computer program (made up of bits) and to develop it together with hundreds or thousands of collaborators. Today, the same can be said for the development of physical products (made up of atoms). The impact of this phenomenon is sure to be significant, leaving behind the notion that manufacturing something has to be particularly expensive and involve assembly lines, heavy machinery,

MANUFACTURING SHARE
OF GDP IN LOCAL CURRENCIES
1970 TO 2010

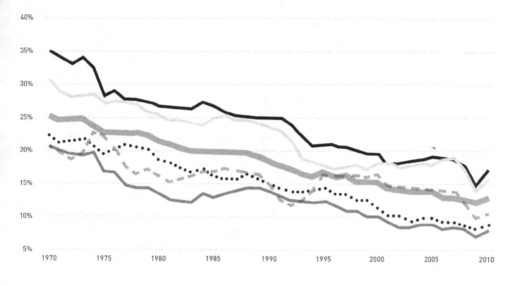

Figure 7.5. Source: United Nations

suppliers, stocks, distribution, logistics, and scalability. Everything had to be carefully planned, and a product's economic feasibility depended on its acceptance by a large number of consumers. But no more.

Makerspaces equipped with 3D scanners, 3D printers, sensors, microcontrollers, and construction and modeling tools are opening the door to huge numbers of people working with this new type of manufacturing. Prototyping, developing, implementing, and testing a new idea is now measured in hours or days—and it is possible to experiment, refine, and collaborate virtually in real time from various points around the world. Technology is now removing the middleman from the production process of physical goods—*this is the disintermediation of manufacturing.*

Understanding maker culture is typically the first step in being able to incorporate it into the production process, followed by the development or use of spaces to facilitate creative activities geared toward solving practical problems. Universities, research centers, and corporations are already busy integrating maker culture into their respective cultures. GE, Intel, Microsoft, and Google are providing spaces for creation that breaks with traditional models in favor of the maker style of addressing problems. And the National Aeronautics and Space Administration (NASA) runs a program called NASA Solve, sharing real problems with the public and rewarding the creators of the winning solutions. In 2009, for example, one of the winners used their own makerspace to develop prototypes for gloves to be used by astronauts, for which they received $200,000 dollars.

VIRTUAL REALITY AND VIDEO GAMES

THE FIRST INDUSTRIAL REVOLUTION, WHICH OCCURRED between the second half of the eighteenth century and the first half of the nineteenth century, ushered in the entrenchment of a new society where science and logical reasoning were gaining increasingly greater relevance. So it is no wonder the appearance of the supernatural in literature declined during this time—and it was precisely because of this decline that English poet Samuel Taylor Coleridge (1772–1834) used the phrase "suspension of disbelief" in his 1817 book *Biographia Literaria*.

THE SUSPENSION OF DISBELIEF

Coleridge coined this term to explain how an enlightened reader should respond to works with fantastical or supernatural elements: They should temporarily suspend their tendency to critique and question, and instead accept whatever they read in the narrative. That's what we do when we consume any type of fiction—whether in books, movies, or any other medium. We simply accept a different reality.

We all create and accept alternative realities on a daily basis—through our dreams. Some don't make much sense, taking place in locations where we have never been or that may not even exist; others are so real that we have difficulty convincing ourselves when we wake up that they were just a dream. The fact is the human brain is predisposed to creating its own reality.

Any method of tricking the mind to make us believe that something unreal is happening before our eyes is the subject of study by neuroscientists who are interested in learning about how our brains process and interpret information. In November 2008, scientists led by Stephen Macknik and Susana Martinez-Conde at the Barrow Neurological Institute in Arizona published a scientific paper entitled "Attention and awareness in stage magic: turning tricks into research." The paper surmised that an excess of stimuli that need to be processed by the brain forces us to take shortcuts, creating a reality that uses simple mental models that leave us easily tricked. (Good magic takes advantage of this process to create illusions.)

Thus, it is not surprising that the desire to create devices to suspend an audience's disbelief is a drive that has occupied writers, scientists, philosophers, engineers, filmmakers, and poets for years. According to the *Online Etymology Dictionary,* compiled by Douglas Harper with the aim of detailing the origin of English words, the use of the word *virtual* as "being something in essence or effect, though not actually or in fact" has been in place since the middle of the fifteenth century. The word comes from the Latin *virtus*, which conveys the idea of excellence or efficacy.

Superimposing virtual elements over real elements (through AR) or immersing the user in a completely 3D environment with images, sounds, and interactivity are efficient ways to deceive the brain. These two techniques merge technology and the study of the cognitive mechanisms we use to perceive the reality around us to create applications for multiple industries.

The foundations for the popularization of applications connected to VR are already installed in our technological infrastructure: a combination of advances in hardware and software development (think VR glasses), the widespread use of smartphones (think *Pokémon Go*), and access to high-speed data networks (think VR broadcasts).

STEREOSCOPES

In the 1830s, English scientist Charles Wheatstone (1802–1875) created the first so-called stereoscopes, effectively a device to transform 2D images into 3D representations of reality. He had noticed that when two pictures designed to simulate left- and right-eye views were presented exclusively to the appropriate eye, the brain created a 3D representation by itself.

Nearly a decade later, Scottish inventor David Brewster (1781–1868), who specialized in optics, produced portable stereoscopes, which, according to many accounts, made an impression on Queen Victoria at the Great Exhibition held in London in 1851. Some of the visitors to this innovations fair were naturalist Charles Darwin, writers Charles Dickens and Charles Dodgson (better known as Lewis Carroll, author of *Alice in Wonderland*), and poet Alfred Tennyson. In 1939, the View-Master was patented and made popular with cardboard disks containing pairs of images (one for each eye), which were interpreted by the brain as a single 3D image.

Figures 8.1 and 8.2. Charles Wheatstone (1802–1875), English scientist and inventor of the stereoscope. His contributions included important advances in studies on the behavior of electricity and telegraphy.
Source: Smithsonian Institution Libraries

Figures 8.3 and 8.4. David Brewster (1781–1868), the scientist whose work was focused on optics and who invented devices such as the portable stereoscope. Source: Wikimedia Commons

One of the most promising and revolutionary areas that will be impacted by VR and AR technologies is professional training. In fact, in 1929, the first flight simulator—one of the training applications with the richest history in the world of computing—was created by American entrepreneur Edwin Link (1904–1981). Called the Link Trainer, this piece of electromechanical equipment consisted of a cockpit connected to engines that responded to the controls and that simulated turbulence. Since the Second World War, it is estimated that more than 500,000 pilots have been trained using some version of this simulator.

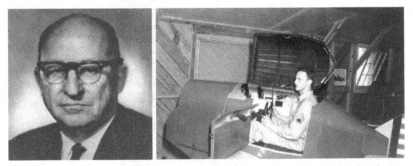

Figures 8.5 and 8.6. Edwin Link (1904–1981), inventor of the flight simulator. Source: Wikimedia Commons

Training professionals to handle complex situations under realistic conditions—while in an airplane, a helicopter, or a spacecraft, dealing with unplanned events during a surgical procedure or a natural disaster, or refining the techniques used for the extraction of natural resources—is not only expensive and inefficient, but also practically unviable.

Rather than only reading about which safety measures to take during an accident on an oil platform or a leak at a nuclear facility, workers can now be trained in realistic settings created by a computer to take the required steps efficiently and intuitively. Because these steps are presented in a simulated setting with detailed objects and correct dimensions, if a dangerous scenario occurs in the real world, the workers' muscle memory will already have been formed to take the correct action in that particular situation.

The first conception of a system in which it was possible for a person to virtually experience situations was recorded by American writer Stanley Weinbaum (1902–1935) in his 1935 book *Pygmalion's Spectacles*. The equipment described in the book accurately anticipates the current structure of VR systems—down to the use of goggles to become immersed in a fictitious 3D environment.

The VR industry owes much to the pioneering work of a filmmaker named Morton Heilig (1926–1997). In the late 1950s he developed Sensorama, nicknamed "cinema of the future" and patented it in 1962. With 3D images, stereo sound, a chair that reacted to events shown on the screen, a device to reproduce specific smells, and even a fan, Sensorama immersed the viewer in one of the movies prepared specially for the equipment. It is considered to be one of the most important breakthroughs in the history of VR.

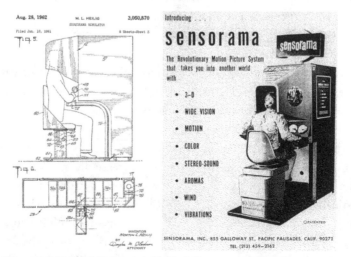

Figures 8.7 and 8.8. Morton Heilig's Sensorama Simulator patent and machine. Source: ResearchGate.net

Work on head-mounted displays (HMDs) started in the 1960s. The HMD's function is to react to head motions made by the wearer, which is a fundamental requirement for the illusion of VR to be understood by the brain, since the images shown change depending on the direction the user is looking. The first HMD—so heavy it needed to be suspended from the ceiling—was developed by a team led by Ivan Sutherland, one of the most important pioneers in computer graphics, who was doing his PhD under Claude Shannon (1916–2001), one of the most important names in computer science.

VIRTUAL SICKNESS

The illusion produced by VR is a direct result of how we perceive the world. VR goggles and headphones are able to trick the brain through sight and hearing (the two primary senses that give us information) so that we have the impression that we're looking at an environment without the limitations of a screen (as in a movie or TV). We have the perception of being immersed in a new world—which may or may not be similar to the reality we know.

Google (Cardboard and Daydream), Samsung (Gear VR), Facebook (through the July 2014 acquisition of Oculus VR for $2 billion), Microsoft (HoloLens), Sony (via Playstation VR), and HTC (Vive) are some of the tech companies offering devices to access this market. Intel is developing technologies to transmit and receive information on wireless HMDs, which is critical to the large-scale adoption of the technology and to allow sporting events and live shows to be watched with 360-degree views or full in-arena experiences using VR.

Despite consistent growth on a yearly basis, the vast majority of VR devices out there are pretty simple and inexpensive, like the Google Cardboard, which is essentially a cardboard box into which a smartphone is inserted horizontally, closing off the user's field of view and enabling their immersion in the setting run by the VR app. With the expansion of content available in VR, more sophisticated equipment can significantly increase its market penetration (assuming that the industry will move toward some type of standardization). IDC estimates that shipments of AR/VR headsets are set to increase roughly eight-fold, from about nine million in 2019 to almost 70 million in 2023.

Large-scale adoption of VR depends both on the hardware (the device used to access the content) and on the software (the content that is experienced in a VR or AR environment)—but it doesn't end there.

One of the industry's biggest challenges is already known: It is called VR sickness, or a feeling of nausea that many people suffer after using VR equipment for over an hour (some people feel it almost instantly). The causes seem to be linked to the same mechanisms that cause kinetosis, or motion sickness, which is a conflict between what the user perceives or captures with their eyes and their perception of movement (captured by the vestibular system, which is responsible for spatial orientation and balance).

The technical challenges that remain for VR to become prevalent seem to indicate a phenomenon that the consulting firm Gartner calls "the trough of disillusionment." After high expectations, practical challenges become apparent, and adjustments must be made to the technology. Two years after acquiring Oculus (a company that makes goggles for VR) for $2 billion, Mark Zuckerberg (founder and CEO of Facebook) stated in a February 2016 interview that he believed the large-scale adoption of VR would take at least 10 years. But several companies have already started to capitalize on the benefits of using VR in medicine, engineering, retail, and real estate as a tool both for training and for entertainment.

PLAYING GAMES

According to industry intelligence firm Digi-Capital, the VR and AR market should exceed $80 billion in revenue by 2023. This growth is expected to occur in no small measure thanks to games (both for consoles and for personal computers) and to headsets for smartphones. Games where the point of view is the player's (first-person shooter games, or FPS) and role-playing games in infinite worlds (where players take on the identity of exotic characters) are the perfect opportunities for the use of VR. Rather than looking at a screen, players are transported to a new world. It is a sensation that can only really be understood by someone who has had the opportunity to experience one of these systems. In a report published in June 2020, Grand View Research estimated that the global VR gaming market was valued at $11.5 billion in 2019 and was expected to grow at a compound annual growth rate of 30.2% until 2027.

Adversaries controlled by the computer with artificial intelligence techniques and the capacity to encounter other players in virtual arenas take the gaming experience to a new level. Many popular titles may be played in this type of setting through the use of specific software drivers, and new games that leverage this technology continue to be launched regularly. The global games industry generated approximately $150 billion in revenue over 2019, according to Newzoo, so any technical innovation that could affect its landscape deserves attention. Content producers (whether for TV, film, or amusement parks) are already positioning themselves for the moment when this content in alternative realities can be consumed by everyone—for now, VR penetration remains relatively modest.

The VCR was an important innovation in the electronics market in the 1980s. After a battle over format between Sony (Betamax) and JVC (Video Home System or VHS) that was won by JVC, millions of consumers purchased devices to watch movies stored on magnetic tapes that were able to store a few hours of video with a quality that would only be considered reasonable by today's standards. Neighborhood movie rental outlets quickly proliferated, paving the way for giants like Blockbuster. After once employing over 80,000 people worldwide, it declared bankruptcy in 2010 due to competition from services such as Netflix. It is a case that calls to mind what happened with Kodak, which was likewise not agile enough to protect itself from the rapid pace that innovation imposes on business. Japanese VCR-maker Funai Corporation made its last batch of the machines in July 2016.

When the VCR gained wide popularity, many said it would be the end of movie theaters. After all, why would someone go all the way to a theater to watch a movie that would end up being available for rent a few months later and then could be watched as many times as the consumer liked? But quite the opposite happened. The film industry continued to grow, incorporating cutting-edge technologies in theaters and consistently improving the experience of going to the movies. Bigger screens, higher resolution, enhanced sound quality, the social aspect of the overall experience—it was difficult to compete with that, as sophisticated as the consumer's home theater system may have been.

With the popularization of VR and AR devices for household and business uses, it is possible we will witness a similar phenomenon in fun centers and amusement parks in general. In Seoul, the Lotte World amusement park is now offering VR goggles for riders on a few of their attractions. If falling from a 70-m (230-ft) tower isn't thrilling enough, the rider can use a VR system in which the illusion of a much bigger fall is created—all while in a futuristic city and being saved by a virtual robot. In Beijing, the SoReal entertainment complex is dedicated exclusively to VR attractions, and one of its founders was Zhang Yimou, the movie director in charge of the opening and closing ceremonies at the 2008 Olympic Games in Beijing. Parks specializing in roller coasters, like Six Flags, are adapting attractions to synchronize the ride with a story in a 3D high-speed setting that transports the rider to another reality. Disney is also leveraging AR and VR both as a way to experience its attractions and to explore new avenues of content creation.

READY GAME ONE

One hundred kilometers (60 mi) from New York City, Brookhaven National Laboratory, founded in 1947, carries out research on nuclear physics, nanomaterials, and energy. American physicist William Higinbotham (1910–1994) headed the lab's instrumentation group from 1951 to 1968. He worked at Los Alamos National Laboratory on the timing device for the atomic bomb during the Second World War. After the war, he was one of the founders of the Federation of American Scientists, whose mission was to make the world safer by encouraging the control of nuclear and biochemical weapons.

Back then, the lab opened its doors every October so visitors could learn about its work. In 1958, Higinbotham thought it would be interesting to create an interactive demonstration for visitors to show how scientific efforts are relevant to society. This was the same year that the Advanced Research Projects Agency (ARPA), now known as DARPA (Defense Advanced Research Projects Agency), was created by the US government as a direct response to the successful launch of the Soviet satellite Sputnik 1 in 1957. The agency was responsible for the eventual creation of the Internet, as mentioned in Chapter 5.

Figures 8.9 and 8.10. William Higinbotham (1910–1994), American physicist and one of the pioneers of video games. Sources: Brookhaven National Laboratory and Dreamstime

Higinbotham's interactive demonstration ended up becoming what many consider to be the world's first video game: *Tennis for Two*. The game used an oscilloscope (used for measuring voltages) that showed the side view of the trajectory of a ball over a tennis court; the trajectory was calculated by a computer with two connected controls designed by Dave Potter and assembled by Robert Dvorak.

FOUR "FIRST VIDEO GAMES"

Four other candidates vied for the title of first video game (but based on the current definition, the title seems to rightfully belong to *Tennis for Two*).

- The first was a cathode-ray tube amusement device, patented in 1947 by Thomas Goldsmith (1910–2009) and Estle Mann. This device is the oldest known interactive electronic game (although it was not run by a computer).

- The second was a tic-tac-toe machine called *Bertie the Brain*, from 1950. It ran on a computer, but provided feedback through lamps, rather than in images that were updated in real time.

- The third and fourth contenders are, respectively, another tic-tac-toe game and a checkers game, both implemented in 1952 by British computer scientist Christopher Strachey (1916–1975). These were simply for academic or demonstration purposes.

Meanwhile, at MIT, employees and students connected to the Tech Model Railroad Club explored the video game possibilities of the PDP-1 microcomputer, which weighed more than 700 kg (1,500 lb). Led by computer scientist Steve Russell, a team created the video game *Spacewar!* and debuted it at the 1962 Science Open House. Six years later, Russell gave programming classes at a school in Seattle. Among his six students were then-teenagers Bill Gates and Paul Allen (1953–2018), who went on to found Microsoft in 1975. *Spacewar!*—adapted and improved by different groups—spread to other universities and demonstrated the cultural (and potential economic) impact that video games were going to have.

COMPUTER SPACE AND *GALAXY GAME*

Two pioneer duos in the expansion of video games are Bill Pitts and Hugh Tuck, and Nolan Bushnell and Ted Dabney (1937–2018). These teams developed the concept of a machine that gives players access to video games after dropping coins into a slot.

In the late 1960s, Bill Pitts (then a student of electrical engineering at Stanford University) and Hugh Tuck (then a student of mechanical engineering at California Polytechnic State University) used to play *Spacewar!* on the PDP-6 installed in what would become the Stanford Artificial Intelligence Lab (SAIL). There they hatched the idea of *Galaxy Game*. With financial support from Tuck's family, the first version of the machine was placed in the student union building at Stanford University in August 1971—for testing.

That's when the paths of the two duos crossed, since Bushnell—who may have had his first contact with *Spacewar!* only after 1969—wanted to compare the designs of *Computer Space* and *Galaxy Game*, which had similar objectives. Electrical engineers Bushnell and Dabney met when working at electronics company Ampex. In 1971, they decided to found a company called Syzygy (which means the nearly straight-line configuration of three celestial bodies), with the aim of creating video games that needed coins to be played.

Bushnell's search for a manufacturer ended in a surprising way. During a dental check-up, he happened to mention his game. The dentist then introduced him to the sales manager of Nutting Associates, founded in 1967, whose main product at the time was a question-and-answer game called *Computer Quiz*. Not only did William Nutting agree to make the game, but he hired Bushnell as the head engineer, paying Syzygy 5% on each machine sold.

Unlike *Galaxy Game*, which remained in the university setting, around 1,500 units of *Computer Space* were sold—enough to convince Bushnell and Dabney to pursue the idea independently. But when they actually left Nutting Associates and decided to finally incorporate their company, they discovered the name Syzygy was already registered. A big fan of the ancient Chinese game *Go*, Bushnell selected a term from the game meaning something like "I'm about to win"—like checkmate in chess—and in 1972, Atari was born.

MAGNAVOX ODYSSEY, *PING-PONG*, AND *PONG*

That same year, the world was introduced to history's first video game console: the Magnavox Odyssey.

Creator Ralph Baer (1922–2014) emigrated with his family from his native Germany to the United States at age 16, following a significant increase in the persecution of Jews. He became a radio repair technician and served in an intelligence unit of the US Army during the Second World War thanks to his fluent German. In 1949, he became one of the first graduates of the American Television Institute of Technology, which landed him a job at the Loral Corporation, a defense equipment supplier that had decided to develop a new television model in 1951.

To test the TVs, Baer used equipment that filled the screen with horizontal and vertical colored lines that could be manipulated, and he suggested that consumers be given access to this pastime, when they no longer wanted to watch the current day's programming. The idea of viewers interacting with their TV was dismissed by Loral's management—but not forgotten by Baer.

In 1966, he became instrument design lead at Sanders Associates. Sanders was a supplier of military equipment to the United States government. It was acquired by Lockheed in 1986 and sold to the British defense multinational BAE Systems in 2000. In his free time, he worked on a secret project with the assistance of two technicians (William Harrison and William Rusch): The LP Channel (for *Let's play*) would result in the world's first household video game console. In 1971, the American electronics company Magnavox (acquired three years later by the Dutch corporation Philips) acquired the rights to the Magnavox Odyssey, and sales began the following year.

Twelve games came preprogrammed into the Odyssey, one of which was *Ping-Pong*, a successor of *Tennis for Two*, and the cornerstone for several lawsuits filed by Baer and by Sanders Associates against imitations. The most famous of these imitations (actually more successful than the original) was *Pong*, developed by Allan Alcorn—the first engineer hired by Atari with the explicit mission of creating an arcade game based on Odyssey's *Ping-Pong*. The commercial success of *Pong* (starting in late 1972) led to *Home Pong*, an Atari console sold in 1975 whose sole purpose was to allow families to play the game at home.

Bushnell sold Atari to Warner Communications (currently AT&T's Warner Media) in 1976 to get the funds he needed to develop a generic console that, unlike the Odyssey, could run any game—not just the ones preloaded onto the console. One year later, Atari released the 2600 console, known as the Video Computer System (VCS), which sold around 30 million units. (The Magnavox Odyssey and its successor, the Odyssey 2, sold around 2.5 million.)

Figure 8.11. Ralph Baer (1922–2014), inventor of the first video game console for household use. Source: Wikimedia Commons

But *Pong*'s story doesn't end there. Two young men worked on the idea (possibly Bushnell's) of developing a one-player version of *Pong*. Their efforts resulted in the game *Breakout*, which was released into arcades in 1976. One of the men was not very technical but was extremely persuasive; the other was incredibly talented in electronic circuits and logic and had worked at Hewlett-Packard on the development of a new scientific calculator. Both were called Steve—one was Jobs (1955–2011), and the other was Wozniak. Together, that same year, they founded Apple Computer Company (currently Apple Inc.).

The video game industry, effectively created in the late 1970s, continues to grow at breakneck speed. Developing new titles costs tens—sometimes hundreds—of millions of dollars, marketing campaigns are comparable to if not larger than Hollywood blockbusters, and professional actors—like Kristen Bell, Mark Hamill, Rami Malek, Ellen Page, Keanu Reeves, and Emma Stone—are paid to have their movements or voices captured in realistic sequences. At the same time, eSports—basically, competitions using video games—around the world are awarding prizes to players of all ages and nationalities, and contracts for broadcasting rights feed a growing global audience: Examples include the Fortnite World Cup, League of Legends World Championship, Rainbow Six Siege Six Invitational, and EVO Championship Series.

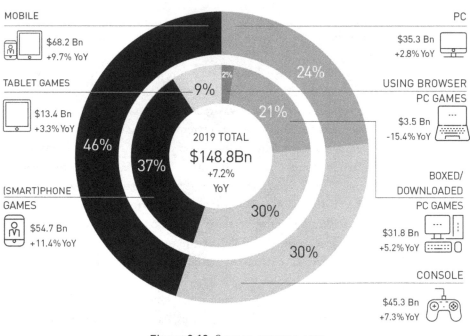

GLOBAL GAMES MARKET—2019
(ESTIMATE, IN DOLLARS)

BY DEVICE TYPE AND SEGMENT,
WITH YEAR-OVER-YEAR (YOY) GROWTH DATA

MOBILE

$68.2 Bn
+9.7% YoY

TABLET GAMES

$13.4 Bn
+3.3% YoY

(SMART)PHONE
GAMES

$54.7 Bn
+11.4% YoY

PC

$35.3 Bn
+2.8% YoY

USING BROWSER
PC GAMES

$3.5 Bn
-15.4% YoY

BOXED/
DOWNLOADED
PC GAMES

$31.8 Bn
+5.2% YoY

CONSOLE

$45.3 Bn
+7.3% YoY

2019 TOTAL
$148.8Bn
+7.2%
YoY

2%
9%
24%
21%
46%
37%
30%
30%

Figure 8.12. Source: newzoo.com

ESPORTS
SOURCES OF REVENUE WITH
YEAR-OVER-YEAR (YOY) VARIANCES

GLOBAL MARKET—2019

(ESTIMATE, IN DOLLARS)

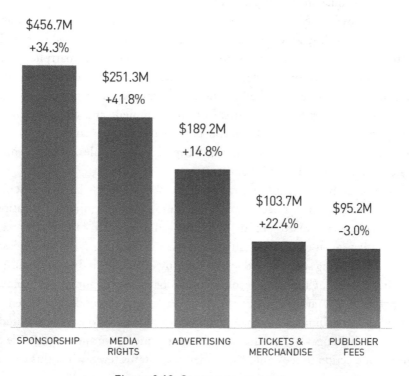

$456.7M
+34.3%

$251.3M
+41.8%

$189.2M
+14.8%

$103.7M
+22.4%

$95.2M
-3.0%

SPONSORSHIP | MEDIA RIGHTS | ADVERTISING | TICKETS & MERCHANDISE | PUBLISHER FEES

Figure 8.13. Source: newzoo.com

AUGMENTED REALITY AND GAMIFICATION

Eran May-raz and Daniel Lazo produced a short film called *Sight* in 2012 for their graduation project at the Bezalel Academy of Arts and Design in Israel. The story shows, in a not-too-distant future, the world as a setting for AR apps, through the use of special contact lenses. In the very first scene, the main character is lying face down on the floor, flying through hoops to score points while exercising. Simple tasks such as slicing a cucumber or scrambling eggs are also gamified, and even dating becomes a points-awarding experience.

The marriage between AR and the concept of gamification is a powerful combination that is set to change the way we learn, study, teach, train, and exercise. Considering the importance most people place on their respective social network accounts and on gaining the largest possible number of followers, contacts, or shares, how our actual lives play out in society, ironically, seems increasingly linked to our digital lives.

Let's now analyze both elements—AR and gamification—separately. Augmented reality, as we have seen, is the technology that allows digital elements to be presented in the real world, whether via a smartphone, tablet, special goggles, or any device with a camera connected to a processor. Gamification can be described as the transformation of a given task into a game that exists *outside* the context of entertainment and competition. For our purposes, let's use the definition put forth by American researcher and designer Jane McGonigal: All games have objectives, rules, and a feedback system—and participation is voluntary.

The objective of gamification is to enhance the user's engagement with a given system or service, aiming at (among other things) operational improvement, increased productivity, and an increased capacity to retain information—you can earn points by exercising, finishing tasks on time, saving money, buying specific products, answering questions in discussion forums, making restaurant and hotel recommendations, or doing homework. Andreas Lieberoth from Aarhus University in Denmark says this concept's efficiency is linked to our willingness to change our behavior if situations are presented as games or challenges. When this happens, our competitive and curious side can spring into action to seek out achievement and status (even when this status is represented by virtual badges, as is often the case).

THE OREGON TRAIL

Early childhood learning has been profoundly influenced by the concept of gamification and its ability to stimulate and boost students' interest in topics like reading, counting, and identifying sounds. Games hold children's attention and increase their focus and the efficiency with which they learn. In fact, the use of computers in education started in the early days of the Digital Age—more precisely, in 1971, with a game developed for 13- and 14-year-old students in Minneapolis, Minnesota, *The Oregon Trail*.

Don Rawitsch, Bill Heinemann, and Paul Dillenberger, three university students from Carleton College near Minneapolis, were slated to teach primary school classes under the supervision of their instructors.

Rawitsch (who was studying history) was assigned to teach eighth-grade students about the settlement of the western United States during the nineteenth century. He imagined that a board game depicting the pioneers' journey could be more interesting than the traditional way of teaching this content, and he started to work on cards that described situations that could be faced by the settlers: illnesses, attacks, hunger, thirst, problems with the wagons, and lack of appropriate clothing.

To be closer to the schools where they taught, the three friends rented an apartment in Minneapolis. Heinemann and Dillenberger—who were studying mathematics and had taken a programming course at Carleton—believed in the teaching potential of computers, even though they did not have any idea of what content to offer. One day, when they arrived back at the apartment, they saw Rawitsch with his cards—and the idea was born immediately.

They had two weeks to create a computer program that could simulate the journey, and that's just what they did. *The Oregon Trail* was one of the pioneers in the genre of edutainment (education + entertainment) and became a standard for American students. The game's objective was to simulate the nearly 3,000-km (1,900-mi) journey by a family of pioneers in 1847, starting in the state of Missouri and ending in Oregon.

The program was loaded onto an HP 2100 that served the region. Students could interact with the program using a teletype that was housed at the school and connected to the server by telephone line.

In a 2017 interview, the creators still remembered the reaction to their work. Students arrived earlier at school and left later to have the opportunity to play as many times as possible. According to the trio, the students organized themselves so that the best typist operated the teletype, the next student followed the progress along the map, and the third managed the budget to ensure there were enough supplies for all settlers.

When the school year ended, the game was deleted from the server—but a copy of the source code was kept by the developers. In 1974, Rawitsch went to work at the Minnesota Educational Computing Consortium (MECC), and (with Heinemann and Dillenberger's consent) he retyped the eight hundred lines on MECC's server, which served the entire state. The game was very popular for several years and was released on multiple platforms, collectively selling more than 65 million copies.

EDUCATION

OUTSOURCING MEMORY

GENERATION Z—WHOSE BIRTH YEARS RANGE FROM THE MID-
1990s to the early 2010s—have enjoyed a world full of technologies that are now well established but were relatively new to previous generations. For Gen Z, smartphones, tablets, Wi-Fi, likes, shares, and real-time information are fundamental pillars of society, and in large measure, are the way in which they interact among themselves and with those around them. Over the years, their trust in and dependence on technology trickled up to previous generations: Millennials (born between 1981 and 1996), Gen X (born between 1965 and 1980), and even, to some degree, baby boomers (born between 1946 and 1964). This dependence is now a significant part of our lives—after all, the smartphone is always within our reach, and therefore no piece of information is out of reach. Simple things like telephone numbers are no longer memorized. In fact, memory has been outsourced—to the machines around us.

This fact has important implications.

In December 2013, Dr. Linda Henkel from the Department of Psychology at Fairfield University in Connecticut published a research paper examining how taking photos affects our memory. She visited a museum with a group of students, asking them to take pictures of random art objects. Her conclusions were counterintuitive: The details and locations of the objects that were *not* photographed were remembered better than those that were. However, if the photograph focused on a particular detail, the object was remembered more vividly. The hypothesis that taking pictures will help us remember events in our lives turns out not to be true.

LEARNING TO LEARN

Integrating technology into the educational process is not a simple matter. When you add into the dynamics of the learning process the fact that computers and smartphones have been in the lives of children from when they were infants, things become even more complicated. The way children learn and retain information changes significantly with the use of technology. Educating is more than simply transmitting knowledge, and some experts in the field argue that the learning process itself is more difficult when technology plays a leading role.

American author Nicholas Carr was a Pulitzer Prize finalist in 2011 for his book *The Shallows: What the Internet Is Doing to Our Brains*, where he states that the experience of being online is like trying to read a book and solve a puzzle at the same time—there is no room for concentration or in-depth perception due to the interruptions caused by the tool itself. His thesis is backed up by studies that indicate that the barrage of information we face from a typical website—with endless links to other websites that include audio, video, comments, and ads—hinders our capacity to process the information since it keeps our brain distracted as it tries to make decisions on what to click on or not. Learning takes place when knowledge being absorbed by the short-term memory is transmitted into long-term memory. The fewer the interruptions and distractions our short-term memory has to deal with, the more efficient this transmission will be. When we are bombarded with excessive stimulation, our capacity to store and connect new knowledge to what we already know is impeded.

In fact, the brains of frequent Internet users—the configuration of their actual neural connections—are different from the brains of people who use the Internet less frequently. Dr. Gary Small, a psychiatry professor at the University of California, Los Angeles, compared magnetic resonance imaging (MRI) data from the brains of volunteers (both Internet veterans and novices) while they performed online searches. The areas of the brain that were activated were significantly different in each group, but just five days after using the Internet for at least one hour a day, the novices' brains were already behaving like the veterans', with additional activity detected in decision-making and complex reasoning areas of the brain. After a relatively short time interacting with the Internet, the dynamics of neuron activation undergo modifications. This needs to be taken into account when attempting to understand the importance and complexity of the evolution of the area of EdTech—or education technology.

The possibilities offered by personal computers for education became apparent early on. Customizing content to the needs of each student and the ease of gathering data to evaluate and assess the learning process contributed to teaching models that can now be easily updated and improved. Any class or course can now be recorded using multimedia resources—images, videos, audio—and it can be made available online to millions of people. Even projects that started out with limited ambitions can become educational phenomena: The Khan Academy, for example, used by tens of millions of students around the world every month, started off as a personal project in 2006 by electrical engineer Salman Khan to help his cousins with their schoolwork.

HOMEWORK

Education sets us apart from other species. Our capacity to index our knowledge, organize it, and transmit it to new generations makes it possible for new humans, just a few years old, to have access to our cultural and intellectual heritage. But despite technological advances in education, much work remains to be done to make this access totally universal and fully realize the transformative potential of knowledge.

PERCENTAGE OF GLOBAL LITERATE
POPULATION AGED 15+ YEARS

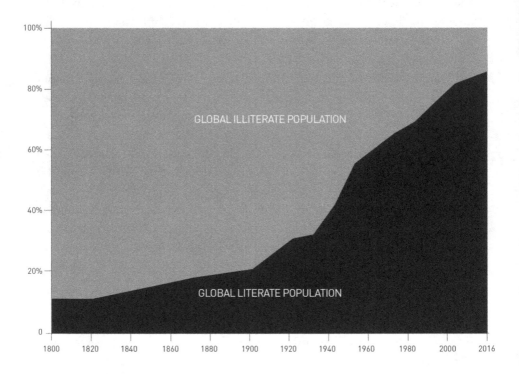

Figure 9.1. In just over 200 years, the world's population increased sevenfold, but illiteracy rates fell from 88% to less than 15%. Sources: ourworldindata.org; Organization for Economic Cooperation and Development (OECD); United Nations Educational, Scientific and Cultural Organization (UNESCO)

The implementation of new educational technologies is often slower than in other industries—not just due to the nature of the activity, but to the complexity of the ecosystem's components, which include students of varying ages and abilities, teachers facing a myriad of unique situations, schools in different geographical locations with different curricula, differing economic conditions, specific regulatory bodies, and government restrictions at the municipal, regional, state, and federal levels.

Schools at all levels—from kindergarten to college and beyond—are going through an important moment of transformation, made painfully clear during the 2020 pandemic. Not only will educational content need to evolve to meet the standards of a post–Fourth Industrial Revolution world, but the way it is transmitted to the students will inevitably change. The number of online courses has been steadily growing with materials from some of the world's most renowned universities: Coursera (École Polytechnique Fédérale de Lausanne, Imperial College London, Johns Hopkins, Princeton, Stanford, the University of Illinois at Urbana-Champaign), edX (Berkeley, Cornell, Dartmouth, ETH Zurich, Harvard, Karolinska Instituet, Kyoto, MIT, Sorbonne Université), and FutureLearn (University of Birmingham, The University of Edinburgh, King's College London, Trinity College Dublin) are companies offering massive open online courses (MOOCs) that rely on unlimited participation and open access via the Internet.

According to global education market intelligence firm HolonIQ, the EdTech market (made up of companies pursuing innovations in the transmission, production, development, and methods of teaching, with the aim of reaching greater numbers of students) is set to reach approximately $340 billion in 2025.

In his 2018 study "Massification of Higher Education Revisited," based on UNESCO (the United Nations Educational, Scientific and Cultural Organization) data, Angel Calderon from RMIT University (Melbourne, Australia) highlighted a nearly 200% growth in higher education enrollments through 2040. Total number of students is expected to rise from about 215 million in 2016 to 380 million in 2030 to almost 600 million in 2040. Pressure for access to quality instruction and knowledge has never been so great, affecting both governments and private groups.

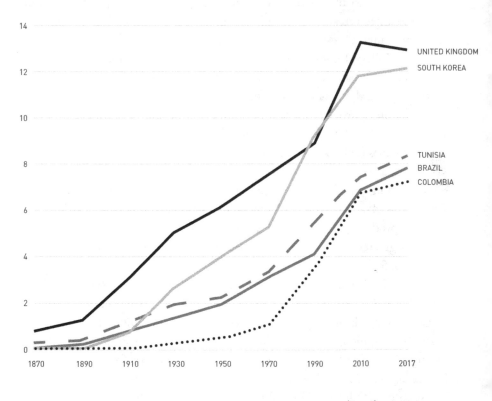

AVERAGE TIME AT SCHOOL (IN YEARS)
ALL EDUCATIONAL LEVELS,
POPULATION AGED 25+

UNITED KINGDOM
SOUTH KOREA
TUNISIA
BRAZIL
COLOMBIA

Figure 9.2 Sources: ourworldindata.org; Lee-Lee (2016);
Barro-Lee (2018); and UNDP HDR (2018)

ESTIMATED PERCENTAGE OF POPULATION
AGED 15+ YEARS WITH A UNIVERSITY DIPLOMA

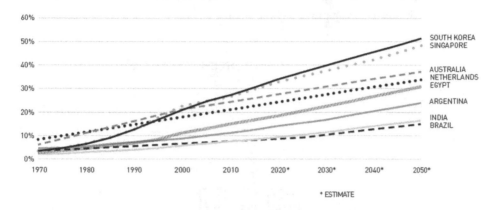

* ESTIMATE

Figure 9.3. Sources: ourworldindata.org; International Institute for Applied Systems Analysis; Lutz, Wolfgang, William P. Butz, and Samir KC, *World Population and Human Capital in the Twenty-First Century*, Oxford University Press (Oxford, 2014)

Kofi Annan (1938–2018), United Nations Secretary-General from 1997 to 2006, stated that "Knowledge is power. Information is liberating. Education is the premise of progress in every society, in every family." In fact, a country's success can be evaluated by looking at the importance its government and its society give to the education of its citizens. It is a long-term investment, but one that, according to scientist, politician, statesman, and diplomat Benjamin Franklin (1706–1790), "pays the best interest."

Educational systems can be evaluated in different ways, but the same countries are usually at the top of the rankings, irrespective of the methodology used: Finland, Singapore, Switzerland, Belgium, Japan, South Korea, Israel, France, the United Kingdom, Denmark, and the United States rank higher or lower depending on the focus of the study (primary, secondary, or post-secondary education) and on the model used in the study. In addition to government investment in education, these countries' public policies are geared toward instruction quality and professional qualification, striving for partisan independence and a consistent method to be followed for decades.

The opportunities and challenges that arise in an increasingly connected and integrated world call into question centuries of educational practices that need to be refined to keep up with the technological evolution that permeates practically all domains of knowledge. Access to information has become virtually universal—whether through searches for specific answers, via quality online courses (which help, for example, in the literacy process and learning basic math), or to earn a university diploma.

KEY SEGMENTS FOR EDTECH COMPANIES

Figure 9.4. Sources: Wikipedia, others

EdTech companies play a critical role amid the structural changes that are underway, seeking the development of tools and/or methodologies that are aligned with the primary lines of educational thinking or expanding the horizons for academic and professional training. As we have seen, the increase in demand for education and the labor market's need for more qualified professionals unquestionably ensure the space for, and importance of, startups built around this concept.

Between 2015 and 2019, according to data gathered by HolonIQ, nearly $28 billion in global venture capital funding was invested in startups from the education sector. Investments are being made in generic online learning platforms, language instruction, early childhood education, technology courses, monitoring and management, personalization of education, exam preparation, and classroom aids, to name just a few examples of this rapidly expanding area. It is interesting to note that EdTech projects do not typically require new equipment or devices, since they rely almost entirely on software. Private educational groups actively seek to invest in new companies that can broaden the reach of their courses to the largest possible number of students, thus enhancing the most fundamental pillar for a country's institutional and social evolution.

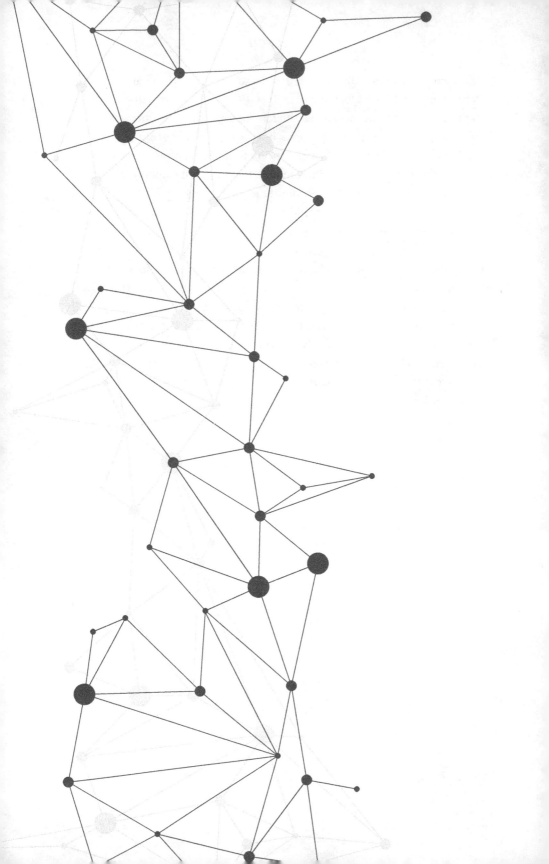

10

SOCIAL NETWORKS

WHAT WILL FUTURE SOCIETY LOOK LIKE AS TECHNOLOGY'S role grows? The answers are speculative and concerning. It is clear that the daily use of sophisticated devices, systems, and equipment will mean significant time savings and increased efficiency, but other consequences will arise that are not yet fully known or understood.

We know the use of technology causes a reorganization of neural connections, a phenomenon known as *neuroplasticity* (the brain's capacity to reorganize itself, transferring processing functions from one region to another, or even strengthening or weakening specific connections). We also are fully aware that the cellphone technology we have in our hands puts us closer to those who are far from us. But it also ends up moving those close to us farther away: Witness tables in restaurants where entire parties are connected to their phones but disconnected from the people in the here and now.

The British TV series *Black Mirror*, created by Charlie Brooker, portrays various scenarios of the near future. Brooker says the series' name comes from the reflection of our images in the screens of our portable devices when they are in off mode. In one episode, *Nosedive*, the socioeconomic situation of individuals is determined by ratings they earn in all of their interactions—in both the virtual and physical worlds. People with higher star ratings get access to better services (more flexible bookings in airlines and better rental cars), shorter lines, and larger discounts, and the episode tells the story of a woman seeking a better social rating who ends up in a downward spiral with dramatic consequences.

Does this seem preposterous? Maybe not. In 2014, the Chinese government launched a program translated as *Social Credit System* (although a more correct translation may have been *Public Reputation System*). The West quickly characterized it as an instrument to be used by the state to exert control over citizens, since the proposition seeks to use the data inevitably made available by the population on a daily basis in their interactions with stores, transportation systems, restaurants, and public services to generate a score that reflects the reputation or trustworthiness (in terms of financial credit, primarily) of each user.

Data that could be used in the calculation of your reputation score in China could include obeying traffic regulations, volunteering for community services, paying bills on time, the quality of your contacts on social networks, your employment relationships, and even your energy savings. A good score would allow you to use services such as renting a car without a deposit, or to avoid lines at the airport.

BELONGING IS BIG MONEY

The way we behave behind our screens reflects old characteristics of humans: the need to belong to a community, to be part of something bigger than ourselves, and to identify with others who have similar interests, principles, or tastes. Until recently, we had to be geographically close to the people with whom we wanted to discuss or share topics of interest. With the Internet and the possibility of instantaneous connections to practically anywhere in the world, this restriction exists no more. It is easy, inexpensive, and stimulating to be part of global communities—so easy that it is unlikely that we would only join one or two. We have dozens of WhatsApp groups, we follow hundreds of people on Twitter and Instagram, we connect with our extended professional network over LinkedIn, we have thousands of friends on Facebook (often people we have never met), and we share information—from the most mundane to the most significant— with a network of contacts made up of family, friends, acquaintances, friends of friends, and acquaintances of acquaintances.

The setup, maintenance, utilization, and expansion of social networks is big business in this new millennium—and the resulting figures are impressive. Facebook (which, as of the end of August 2020, had a market value of around $835 billion and was one of the most valuable brands in the world) had fewer than 200 million users in the first quarter of 2009. Ten years later, this number topped two billion. YouTube—created in February 2005 and acquired by Google in November 2006 for $1.65 billion—has over two billion users. This is also the estimated number of users of the messaging app WhatsApp, which was founded in 2009 and acquired by Facebook in February 2014 for no less than $19.3 billion. Instagram, founded in 2010, was also acquired by Facebook (in April 2012, for $1 billion); it hit one billion users in 2018. LinkedIn, a professional networking service acquired by Microsoft in 2016 for $26.2 billion, had an estimated 706 million users in 200 countries in 2020. And social networks that originated in China also rack up hundreds of millions of users, like WeChat (1.2 billion in June 2020) and TikTok (800 million in July 2020).

Analyzing the phenomenon of social networks is an exercise that cannot be done without paying attention to human psychology. According to the "Global Digital Report 2019," prepared by We Are Social (a global socially led creative agency based in England) and Hootsuite (a Canadian company that develops social media management systems), of the approximately 4.4 billion Internet users, almost 3.5 billion belong to some type of social network. Dr. Ciarán Mc Mahon, with credentials from University College Dublin, is one of the researchers interested in the question "What drives so many of us to participate in these digital organizations?"

One of the basic (and intuitive) reasons for our interest in social networks is the constant flow of content: news from the world, our country, our city, our family; curiosities; photos; videos; gossip. There's a lot to talk about—and it is all new. In the Internet environment, where so much is competing for our time and attention, the element of newness is a significant competitive advantage.

The second major motivation for the popularity of online social communities is much less obvious and connected to an aspect of human behavior identified and studied by American psychologist B.F. Skinner (1904–1990): the effect of reinforcement, whether positive or negative. Actions with negative consequences for an individual are *not* repeated, while those with positive consequences typically are. But in their 1957 book entitled *Schedules of Reinforcement*, Skinner and his colleague Charles Ferster (1922–1981) showed that actions that can lead to *either* negative or positive feedback are almost certain to be repeated. And that is what ends up happening when we participate in a social network: Sometimes we can be rewarded with likes, which were shown to impact teens' brains in a way similar to eating chocolate or winning money, according to a 2016 study by UCLA researchers published in a journal of the Association for Psychological Science. Other times we earn followers, who can be acquaintances or strangers; sometimes we can be ignored—or criticized. Since we don't know what type of reinforcement we'll receive next, we maintain the same pattern of behavior—we keep right on posting our photos, comments, and opinions. The troubling impact of social networking effects in society and the role played by leading tech companies in this process was the focus of the 2020 American documentary and drama *The Social Dilemma*, directed by Jeff Orlowski.

If and when our digital presence is reasonably well established, we start checking our smartphone dozens of times a day, looking for that red dot over the icon of each of our social networks, indicating something new awaits us. This pattern of constant interruptions over the course of our day is cause for concern in schools and in workplaces precisely because interruptions are a detriment to the learning process, concentration, and focus. With these interruptions, drops in productivity and progress are inevitable.

LET'S GO SHOPPING

Before e-commerce (online shopping), a consumer who wanted a certain item had two options: go to where this item was in person or order it by mail from a catalog or via forms in printed magazines or newspapers. But that all changed after what is considered to be the first deal facilitated by the Internet, made in the early 1970s using ARPANET (the precursor to the current Internet) when students at Stanford and MIT organized a marijuana sale.

In May 1984, the first B2C (meaning "business-to-consumer") transaction was made when Mrs. Jane Snowball, 72, a resident of Gateshead, in northern England, purchased margarine, cereal, and eggs using her TV's remote control and a system called Videotex. The device used the principles of e-communication conceived by entrepreneur Michael Aldrich (1941–2014) in the late 1970s, which turned TVs into communication terminals capable of transmitting orders to stores.

Figure 10.1. Michael Aldrich (1941–2014), the English inventor who created e-commerce in the pre-Internet era. Source: Wikimedia Commons

A decade later, in 1994, Jeff Bezos, a former Wall Street analyst, saw a major change around the corner and moved from New York to Seattle, where he founded the company that would become synonymous with e-commerce. Amazon went public in May 1997, and in just 20 years it became one of the most valuable companies (and brands) in the world: In 2019 its revenue was nearly $281 billion, and its market value at the end of August 2020 was around $1.7 trillion. Just like Google, which started off as a search engine and gradually broadened its horizons, today Amazon is much more than just a shopping website, with an important foothold in cloud computing, digital streaming, and artificial intelligence.

The global ascent of e-commerce has brought about significant changes for retailers in all sectors. Initially, the products sold on the Internet were items such as books. The pioneer of this sector was Charles Stack, who in 1992 opened his virtual bookstore, called Book Stacks Unlimited, based in Cleveland, with credit cards as a payment method. Prior to the popularization of the concept of hypertext created by Tim Berners-Lee in 1989 at CERN (the European Organization for Nuclear Research), which eventually became what we know today as the World Wide Web, Stack decided to launch his bookstore into the virtual world when it was still organized around so-called BBSs (Bulletin Board Systems). Using a terminal, a modem, and a dial-up line, users would connect to a server where they could read news, download programs, or exchange messages (any similarity to the Internet we know today is no mere coincidence).

In a 2002 interview with Dustin Klein of the Smart Business Network, Charles Stack spoke about his first online sale, made around a week after he had placed ads about his bookstore in magazines geared toward BBS users (at the height of BBSs, in 1994, their numbers were estimated at around 17 million in the United States alone). His modem was contacted by the modem of a client, and the connection was established. His entire staff gathered around the monitor to watch the historic transaction, which was happening much slower than everyone was expecting.

Frustrated with the delay, Stack interrupted the session with a message to his client: "Why do you like this service?" Another long pause, and, slowly, the initial characters of the response came through: "A . . . s . . . a . . . b . . . l . . . i . . . n . . . d . . . p . . . e . . . r . . . s . . . o . . . n. . . ." The first purchase (probably a gift for someone) was being made by a visually impaired person, who used a voice synthesizer to read out the text from the BBS screen and a Braille keyboard to input information, which accounted for the delay.

In 1996, Book Stacks Unlimited was purchased by Hospitality Franchise Systems for around $4 million (around $6.5 million in today's terms) and, sometime later, it was acquired by American bookstore Barnes & Noble.

With the popularization of the World Wide Web and the creation of browsing platforms, the Internet saw an unprecedented increase in the number of users and its popularity. E-commerce websites multiplied and specialized. It is now possible to purchase practically anything online: from cars to clothing, luxury items to apartments, construction material to computers. According to market research company eMarketer, in 2019 global e-commerce sales totaled about $3.5 trillion, accounting for 14% of overall retail sales—an amount that is set to reach $6.5 trillion by 2023, taking e-commerce's share in retail sales to about 22%.

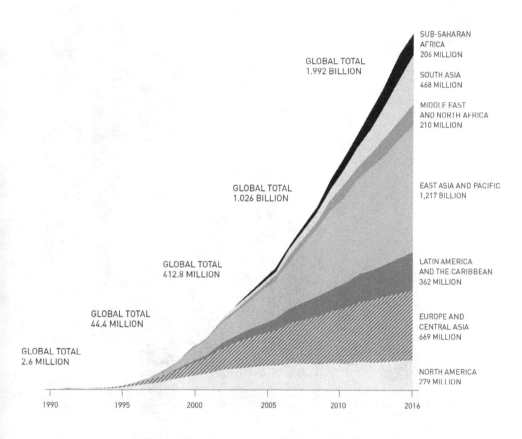

INTERNET USERS BY REGION
1990 TO 2016

SUB-SAHARAN
AFRICA
206 MILLION

SOUTH ASIA
468 MILLION

MIDDLE EAST
AND NORTH AFRICA
210 MILLION

GLOBAL TOTAL
1.992 BILLION

EAST ASIA AND PACIFIC
1,217 BILLION

GLOBAL TOTAL
1.026 BILLION

LATIN AMERICA
AND THE CARIBBEAN
362 MILLION

GLOBAL TOTAL
412.8 MILLION

EUROPE AND
CENTRAL ASIA
669 MILLION

GLOBAL TOTAL
44.4 MILLION

GLOBAL TOTAL
2.6 MILLION

NORTH AMERICA
279 MILLION

1990   1995   2000   2005   2010   2016

Figure 10.2. Source: United Nations, via the
International Telecommunication Union (ITU)

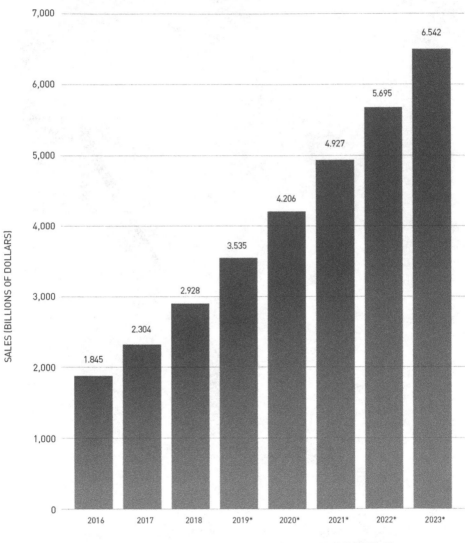

E-COMMERCE: GLOBAL SALES (RETAIL)
2016 TO 2023

*ESTIMATE PRE-COVID-19

Figure 10.3. Source: United Nations, via the
International Telecommunication Union (ITU)

HOW MUCH IS A CLICK REALLY WORTH?

The success or failure of any new network or community is inextricably linked to the number of our peers who are (or are not) engaged in it—and it is possible to measure the associated economic value. About 40 years ago, American electrical engineer and PhD in computer science Robert Metcalfe formulated a principle that became known as Metcalfe's Law.

Born in 1946 and one of the cocreators of Ethernet (the standard protocol for computer networks around the world), Metcalfe postulated that the effects of a network are proportional to the square of the number of elements connected to it—which would indicate, in theory, a measure of the value each network has. This is because, for each new element, the possibilities of connections grow quadratically (more precisely, they increase according to the equation $n*(n-1)/2$, where n is the number of elements in the network). For example, if there are two people on a network, the maximum number of connections between them is just one. With five people, this number grows to 10. With 10 people, it jumps to 45. With 20 people—190 connections.

But could it be that an increase in the number of connections equates to a proportionally quadratic increase in business (that is, the economic value of the network), as suggested by Metcalfe? Several researchers argue that because not all newcomers to a network add the same value, the economic effect of this network is therefore not proportional to the square of the number of users, but only to the logarithm of this value (therefore resulting in a much lower valuation). In a 2015 paper published in the *Journal of Computer Science and Technology*, researchers Xing-Zhou Zhang, Jing-Jie Liu, and Zhi-Wei Xu, from the Chinese Academy of Sciences, analyzed data from two of the world's largest social networks: Tencent (China) and Facebook. They concluded that a network's value does, in fact, grow proportionally to the square of the number of users—corroborating Metcalfe's Law. But the topic is still up for debate, and aspects such as the cost of the network and the value of new users are considered when refining pricing techniques.

In 2000, the number of Internet users was around 410 million; just 19 years later, more than four billion of us were connected. Furthermore, according to web analysis company StatCounter, October of 2016 marked the first time in history that mobile devices outpaced the use of desktop computers for access to the Internet globally. In a few emerging countries (such as Kenya, Nigeria, and India), access via mobile devices accounts for more than 75% of connections.

INTERNET ACCESS
BY DEVICE TYPE
AUGUST 2010–AUGUST 2020

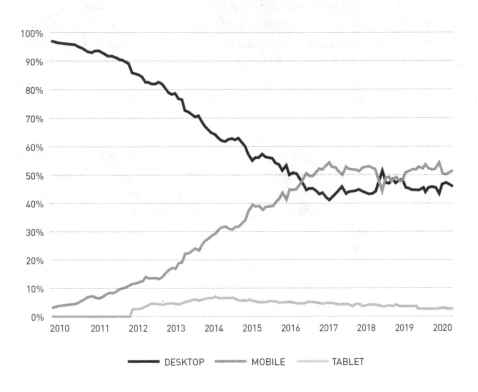

Figure 10.4. Source: StatCounter

But how can the business model of social networks be lucrative, given that they don't usually charge end users a sign-up or monthly fee? How can we measure the profitability of a business that uses a model that doesn't seem to include payments from the end user to the service provider? Let's try to analyze this based on a common indicator in the business world: return on investment, or ROI, which calculates the financial return of an investment.

The ROI obtained through social networks would be equivalent to a company's return after allocating resources (marketing professionals and technicians, for example) to a project. The challenge lies in comparing the money invested against the financial value of the interactions that clients, or potential clients, make via social networks such as Twitter, Facebook, Instagram,

LinkedIn, and YouTube. The relevance of this topic has become so great that a new career has emerged: that of *digital marketing specialist*, who is a professional trained in techniques for measuring and broadening the impact of a brand's online presence.

Two of these professionals are Kevan Lee (marketing director for the social media platform Buffer) and Neil Patel (of Crazy Egg and Neil Patel Digital). Monitoring the effectiveness of campaigns that use social media is a fundamental part of their work, and of the work of their counterparts in many other companies. There are many ways to do this, but often the models consist of defining one or more targets, monitoring interactions on the social networks, and assigning a monetary value to these interactions.

The first step, wherein the targets are set, determines what is relevant to a campaign in terms of business strategy: clicks on a link, gaining new followers, selling specific products, downloads of a file, watching a video, filling out a form—in sum, any type of relevant indicator that can be quantified. This quantification is exactly what happens in the second step. Using tools like Google Marketing Platform (previously known as Google Analytics), Hootsuite, Talkwalker, Brandwatch, and others, it is possible to monitor events such as clicks, likes (when available), and downloads.

The third step—assigning a financial value to the events of interest—may be done based on the so-called lifetime value (LTV) of the client (i.e., how much each client generates in terms of results for the company). For example, in simplified terms, if each client's LTV is $100, and if 1 in every 20 visitors to the website effectively clicks on a specific link, this click is worth $5 ($100 divided by 20 visitors). The result will be the sum of all clicks on all social networks where the campaign is being run. Dividing this value by the investment—planning, developing the system, traditional marketing—you will have the ROI of your online campaign broken out per social network, allowing you to determine which channels are most efficient and attractive for your audience.

LOOKING FOR LOVE

Looking for companionship is human nature, and using technology to expand the opportunities of being part of a community is what online communities are all about. The logical extension of this concept is the search not only for a community but also for *someone* special. Numerous online options allow people to create lasting (or not-so-lasting) connections irrespective of sexual orientation, gender, age, race, or religion.

The first documented use of computers in matchmaking dates to 1959. Two students of Stanford University professor Jack Herriot (1916–2003) submitted the *Happy Families Planning Service* as their final project on the theory and operation of computational machines. Philip Fialer (1938–2013) and James Harvey drew up a questionnaire that was given to 98 single people (49 men and 49 women). Using the university's IBM 650—likely the first mass-produced computer—the students calculated the differences between the responses and assembled a compatibility list of each of the possible pairs.

A few years later, in 1964, Englishwoman Joan Ball (one of the first female entrepreneurs to work in the technology segment) created the St. James Computer Dating Service, the world's first commercial computer-based service to facilitate match-ups. It later became Com-Pat (a play on the words *computer* and *compatibility*) and eventually led to close to 15,000 marriages.

The company was acquired in 1974 by the English service Dateline, founded by John Patterson (1945–1997) in 1966. (After 32 years of operation the service was purchased in 1998 by the Columbus Publishing Group.)

The inspiration for the formation of Dateline came from a visit Patterson had paid to Harvard University. It was there that he learned of the service that eventually led to the formation, in 1965, of the company Compatibility Research, by Jeffrey Tarr, David Crump, and Douglas Ginsburg. People interested in the new service, dubbed Operation Match, mailed in a completed questionnaire and a $3 fee (around $25 in today's terms), and approximately 10 days later they received a computer-generated list with the names and telephone numbers of the most promising candidates according to the compatibility algorithm used. Like Facebook, the initial target market for these services were also students at some of the most renowned universities in the northeast of the United States.

Compatibility Research attracted over a hundred thousand clients in less than a year, and soon other similar services (such as Contact, founded by MIT's David Dewan) emerged. It is important to note that these services were only possible through the rental of the processing power of large machines (starting in the second half of the 1960s) because having a personal computer was unheard of.

Not all such endeavors were scientific: Project Flame was developed by a student at Indiana University and was the subject of a 2014 Slate.com article. Creator Ted Sutton solicited and received completed questionnaires (along with a $1 fee, today worth around $8) from hopeful clients and then simply matched up the questionnaires randomly.

The first matchmaking services were launched in the 1960s, but 10 years later, in the (less innocent) 1970s, the darker and more dangerous side of this type of service came to light. Nathan Ensmenger, a professor at the Indiana University School of Informatics and Computing, recounted the case of a clerk at a bookstore in Times Square, in Manhattan, who sold the addresses and phone numbers of women who used dating services, obviously without their consent.

The risks of dating sites have come to light on more than one occasion, when users post false personal information to attract and assault victims, for example. Hackers also take advantage of the fact that sensitive information is often available in these websites' databases. In August 2015, the personal data of 37 million users of the Ashley Madison website (founded in 2002 for married people looking to have one or more extramarital relationships) was leaked.

LARGE BRAIN, LARGE GROUP

Our need for contact with other humans is a defining characteristic of our species: We are social animals. The theory of evolution developed by English naturalist Charles Darwin (1809–1882) stated that individuals in groups were more likely to survive because the group provided greater access to water and food and protection from predators. Some researchers believe that it was around three hundred thousand years ago—with *Homo sapiens*—that language was born, through the coordination of efforts among the group, using sounds and gestures.

Robin Dunbar, an English anthropologist specializing in the study of primates, proposes that the evolution of intelligence was predominantly an evolutionary response to the challenges of living in society. Dunbar established a metric, known as Dunbar's Number, that expresses the maximum number of stable relationships an individual is capable of maintaining. Dunbar also found a correlation between primate brain size and the size of their social groups: the larger the brain, the larger the group. For humans, Dunbar's Number is approximately 150, indicating that each of us is able to sustain a socially active circle of 150

continued

people (I'm guessing this is far less than your connections on Facebook, WhatsApp, LinkedIn, Instagram, Twitter, and other social networks combined).

How did the creation of communities by primates occur tens of millions of years ago? In a 2011 *Nature* article, Susanne Shultz, Christopher Opie, and Quentin Atkinson, from the Institute of Cognitive & Evolutionary Anthropology at Oxford University, suggested that contrary to the so-called gradualist model (where couples evolved into clans, and then into larger communities) primate societies actually formed quite quickly. According to their research, around 52 million years ago the ancestors of monkeys and lemurs branched out, becoming more active during the day and seeking, through the evolutionary mechanism of survival, safety in numbers: the larger the group, the less vulnerable they were to attacks, and at the same time, more members of the group were available to look for water, food, and shelter.

Now, thousands of years later, technology has become one of the primary tools for sifting through the large numbers of individuals present in our society to help in the selection of a group and a partner.

Match.com, the first dating website in the commercial age of the Internet, was launched in 1995, and it is still in operation (after several changes of control and strategy). In 2019, Match Group (its parent company) had more than nine million subscribers to its several online dating services. The websites and services that followed Match have played huge roles in modifying the social behavior of several generations.

One of the best-known spinoffs of the Match universe is the app Tinder, launched in 2012, and which is based on the geographic location of its users. In just two years, nearly a billion daily swipes were made by its users, resulting in no less than 12 million dates. As is the case with so many other applications of technology, this segment has also generated lucrative business opportunities. The Match Group—which, in addition to Tinder, operates OkCupid—held its initial public offering in November 2015, and it was valued at about $24 billion in July 2020.

In a 2017 article, "The Strength of Absent Ties: Social Integration via Online Dating," researchers Josué Ortega (University of Essex) and Philipp Hergovich

(University of Vienna), revealed how the popularization of these dating sites is facilitating marriages between people of different races. They also postulate that the marriages of people who meet online have greater chances of being successful. Although it is still early to confirm or refute this notion, one of the consequences of this phenomenon is undoubtedly the expansion of the social groups within which we move.

OPINIONS VS. FACTS

When we join a social network, we expect to obtain benefits, whether personally, professionally, or both, and these benefits are directly linked to the number of connections we expect to make. Furthermore, if everyone we know is part of a certain virtual community, we are strongly led to join the same one. But why is this?

The social impulse that makes people follow or mimic the majority has been the subject of academic studies for quite a while. Polish psychologist Solomon Asch (1907–1996) carried out fundamental research on the subject, demonstrating how the behavior of a group significantly influences the behavior of an individual. One of his most famous experiments was carried out in an elevator as part of a sketch for the American TV show *Candid Camera*. In an episode aired in 1962, several actors get into an elevator and stand with their backs to the door, looking at the elevator's back wall. Another person gets in, and after seeing the behavior of his fellow riders, ends up turning to face the back as well—without even knowing why, and challenging his common sense. The experiment is repeated several times, each time with the same result.

Conforming and succumbing to group thinking was also tested in another experiment in the 1950s, where Asch showed that when individuals are pressured by their peers, their opinions are not only influenced, but they start discounting basic evidence. The test was simple: A group of six people (five of whom were actors) were shown a piece of paper with a line drawn on it. Next to the line were three other lines, and each participant had to say which of the three lines had the same length as the original line. The only non-actor participant in the experiment was the last or second-to-last one to respond, after hearing the answers of the others. In most of the cases, when noticing that the entire group gave a certain answer, even if it was clearly wrong, the participant preferred to go along with the group, to avoid embarrassment or drawing attention. When at least one of the actors gave the right answer, however, the

HOW THEY MET

HETEROSEXUAL COUPLES

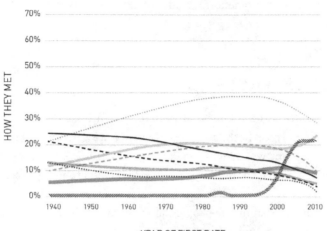

YEAR OF FIRST DATE

HOMOSEXUAL COUPLES

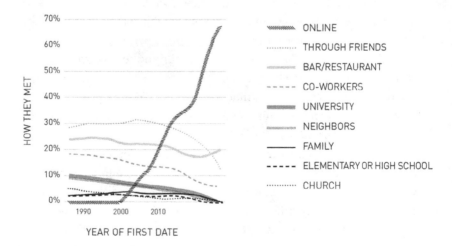

YEAR OF FIRST DATE

Figure 10.5. Source: Michael Rosenfeld and Reuben Thomas,
"Searching for a mate: The rise of the Internet as a social intermediary,"
American Sociological Review, 77(4): 523–547 (2012)

real participant would typically use their common sense and choose not to go along with the majority.

As social networks have become a constant in the lives of the vast majority of the population with access to the Internet, conformity has taken on new dimensions. In this alternate reality, opinions have become more important than facts, and the quick, easy, simple—and often wrong—explanation replaces fact-checking, truth-seeking, science-based facts. If the opinion maker is powerful enough or popular enough, then why bother checking his or her facts? If I can't see the Earth's curvature, I guess it is flat. If someone said that vaccines cause autism, I will not vaccinate my kids. If I simply believe against scientific evidence that a specific medicine should work to treat a deadly virus, then I guess that's ok. Well—it isn't. The powerful Internet stage, which provides a platform to so many to express their opinions and thoughts, also allows for misinformation and ignorance to run amok, and now we live in a dystopian world where *fake news*, *alternate facts*, and *post-truth* are actually things we have to worry about.

In this world driven by social media, the number of likes has become a challenge on sites such as Facebook and Instagram: Likes are an important metric for so-called digital influencers (who are paid by major brands to post images of themselves using their products), but for most users, getting tons of likes has become an at-times unbearable source of pressure. Both Facebook and its subsidiary, Instagram, are developing tests to restrict access to the exact number of likes, an endeavor aimed at improving user experience.

Furthermore, an increase in suicide rates among children, teens, and young adults seems to be an equally disturbing consequence of the excessive use of social media. Data from the US Centers for Disease Control and Prevention (CDC) shows that between 2006 and 2016 there was a more than 70% jump in suicide among youth aged 10 to 17. Also according to the CDC, between 2009 and 2017, there was a 25% increase in the number of high school students who considered taking their own life. Globally, according to the WHO, in 2016, suicide was the second-highest cause of death among people between 15 and 29 (after traffic accidents).

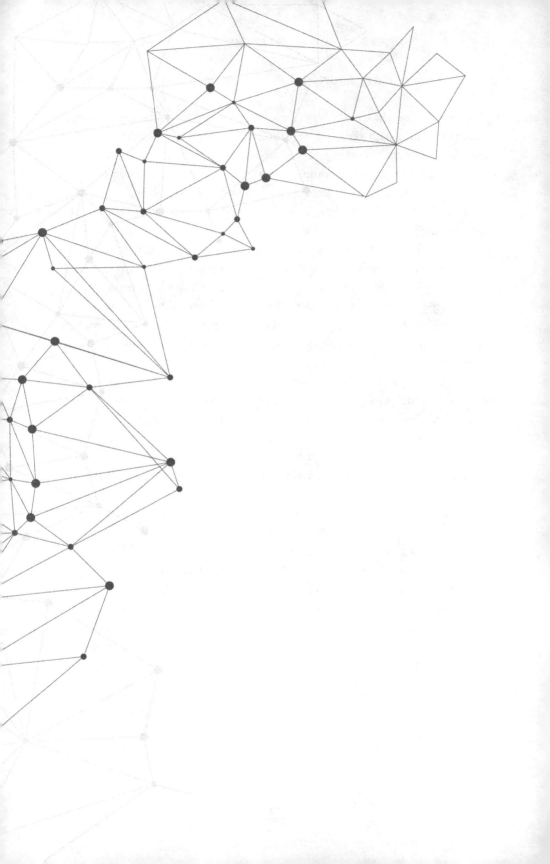

11

FINTECH AND CRYPTOCURRENCIES

THE EVOLUTION OF MONEY

THE CONCEPT OF MONEY HAS BEEN PART OF THE HISTORY OF
civilization ever since humans started to organize themselves around agriculture and livestock raising, about 10,000 years ago. Back then, the trade currency was the object of the trade itself and not an abstraction to which a certain value was attributed (such as a $20 bill). Aristotle (384–322 BCE), the Greek philosopher who also wrote about the nature of work and employment, said that each object could have two uses: its original use and its use in a sale or trade.

Portable objects with intrinsic value became candidates for the role of currency, and ancient civilizations around the world used things like gold, silver, copper, salt, tea, and shells with the specific purpose of facilitating the exchange of goods. The origin of the word *money* itself is probably connected to the temple of Juno Moneta, the goddess and protectress of funds, where the mint of ancient Rome was located. The word *denarius*, in Latin, referred to a type of silver commonly used in Rome both during the Republic (509–27 BCE) and the Empire, before the East/West division (27 BCE–395 CE).

Historians and archeologists believe that the world's first coin—with the same characteristics and symbolism still prevalent today—appeared between 650 and 600 BCE, in Lydia (modern-day Turkey). Strategically located at the confluence of the commercial routes between Asia and Europe, the Lydians were, according to Greek historian Herodotus (484–425 BCE), born merchants. The old expression "to be as rich as Croesus" refers to Lydian king Croesus (595–546 BCE), son of Alyattes (640–560 BCE), both of whom reigned during the height of the region's power. The stories of the place's riches are recounted in Greek mythology: It was in the Pactolus River, which ran through Lydian territory, that King Midas, of neighboring Phrygia, bathed to rid himself of his golden touch, which turned everything he touched—including his food and loved ones—into gold.

The use of paper as currency originated in China, during the Song dynasty (960–1279). Known as *jiaozi,* the bills could be exchanged for coins and traded among individuals. European explorers such as Marco Polo (1254–1324) brought the concept to Europe, but it wasn't until the middle of the seventeenth century that bills started to become popular (likely due to the high rate of inflation caused by the inflow of gold and silver from Spanish colonies).

The first issue of banknotes in Europe was done in 1661 by Stockholms Banco, founded in 1657 and managed by Johan Palmstruch (1611–1671). To bridge the gap between (normally long-term) loans granted to their clients and the sums deposited (normally for a short term), Palmstruch created the *Kreditivsedlar* (credit paper), which were credit notes that could be exchanged for coins. The idea was very well received, but uncontrolled issuances led to the institution's bankruptcy in 1668. In 1695, the Bank of England was founded and started issuing its banknotes.

Banking activity has kept pace with the development of trade since ancient times, and trade as we know it dates from early fifteenth-century Italy, which was the financial center of a world just starting to enter the Renaissance. The use of the word *bank* to describe financial activities probably originated at that time, because transactions were done on *bancas*—the old Italian word meaning "table."

TAKING IT FROM THE BANK

In his 2008 book, *The Ascent of Money: A Financial History of the World,* British historian Niall Ferguson (Harvard University) argues that the concepts of

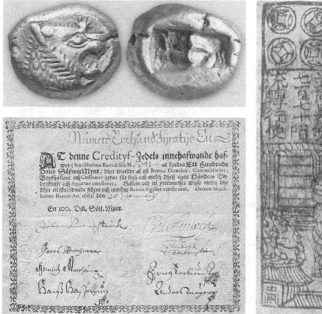

Figures 11.1, 11.2, and 11.3. Clockwise from top-left: this is probably the world's first coin (4.71 g, 13x10x4 mm), minted by King Alyattes of Lydia between 610 and 600 BCE; the jiaozi, a promissory note considered to be the first paper currency in history; and a banknote from April 17, 1666, issued by Stockholms Banco and signed by Johan Palmstruch. Sources: Classical Numismatic Group, Inc., Wikimedia Commons and Wolfram Weimer, Geschichte des Geldes, Insel-Verlag, Frankfurt 1997

credit and *debt* were as important for the evolution of civilization as any key technological innovation. Until recently the intersection between these two areas—technology and finance—has lived far from public view, with the use of mathematical models and computers to price and trade assets around the world, or the development of real-time control and monitoring systems for the activities of banks and stock exchanges. But so-called *FinTech* companies—which merge finance and technology—are rapidly changing this and delivering innovations to the day-to-day workings of the financial system, with crowdfunding platforms, algorithm-based asset recommendations, budgeting apps, and mobile payment systems.

The elimination of intermediaries in business is one of the clear trends of the post–Fourth Industrial Revolution world. As we have discussed, the modern consumer has realized they don't always need brokers or sales reps to perform their transactions. They can purchase goods, obtain services, rent property, buy a car, and hire employees for temporary or permanent positions.

Industries that work almost exclusively with information—such as publishing, education, and financial markets—are areas of great focus for innovators and entrepreneurs. New processes, methods, or models will flourish in brand-new business settings where mobility, connectivity, processing power, and artificial intelligence techniques have become economically viable and accessible. You can self-publish, you can learn at your own time and pace, and you can go about your financial life with more autonomy and freedom.

The financial industry, in particular, is in a unique position. On one hand, it is subject to rigid regulation and constant oversight on a global scale. On the other, it has all the desirable elements for entrepreneurs: scale, antiquated processes, inefficiencies, and (often dissatisfied) clients. According to KPMG's "Pulse of Fintech" report, just between 2017 and 2019 almost $118 billion was invested in startups in this sector. If private equity and mergers and acquisitions activity is also considered, the total value soars to $361 billion.

Part of the explanation for this huge interest in FinTech companies is due to the industry's broad scope (after all, no business can survive without a financial department) and new structural ways companies are set up and managed. In the past the only place to raise funds was at the bank, but today crowdfunding sites abound. The entrepreneur presents their idea to the public via the Internet, specifies how much money they need to start their company, and individuals decide whether to contribute to the project. Kickstarter, Indiegogo, and AngelList are a few examples—and are indicative of an area of FinTech that does away with the need for bank loans.

As could be expected, loans are one of the primary needs addressed by the startups that have decided to operate in the sector. The traditional model for granting credit by a financial institution involves detailed analyses, extensive documentation, and the provision of guarantees so the bank can avoid losses in the event the borrower cannot pay back the loan. Evidently, this process requires time and resources, and it is often not economically feasible in the case of small sums. After the 2008 financial crisis, which originated in the real estate securities market in the United States, it became even more difficult to get a bank loan. For startups from the sector, however, small loans are sometimes processed and granted within minutes, without even speaking to anyone—the app developed for this takes care of everything, carrying out online checks and connecting borrowers and lenders at a competitive cost.

GLOBALLY, 1.7 BILLION ADULTS
DO NOT HAVE A BANK ACCOUNT

ADULTS WITH NO BANK ACCOUNT, 2017

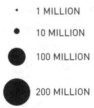

- · 1 MILLION
- ● 10 MILLION
- ⬤ 100 MILLION
- ⬤ 200 MILLION

NOTE: INFORMATION IS NOT SHOWN FOR ECONOMIES WHERE 5% OR
LESS OF THE ADULT POPULATION DOES NOT HAVE A BANK ACCOUNT

Figure 11.4. Source: World Bank, Global Findex database

According to Grand View Research, the digital lending platform market is expected to register a compounded annual growth rate of almost 21% between 2019 and 2026, when just the platforms alone reach valuations north of $15 billion. The sector now includes names that move billions of dollars every year in loans for individuals, small and medium businesses, autos, students, credit cards, and real estate.

The methods used for credit analysis—a fundamental part of the process of granting loans—are also changing. Historically, a client's banking history was used along with their company's financial information, but many FinTech companies seek to serve people with no background in the financial market—a contingent of around 1.7 billion people in 2017 (Figure 11.4)—who usually need small amounts for personal or professional use. The systems for assessing the risk of a given loan applicant include the use of artificial intelligence techniques to analyze, with the applicants' consent, their digital footprint—their profiles on social media, geolocation, browser history, and installed apps. The algorithms will then evaluate all this information and come up with a credit score on the fly that is used to approve (or not) the loan.

The processes for obtaining loans are not the only things being modified: The act of sending and receiving money is a focus of innovation by FinTech companies. One of the best-known examples is PayPal, created out of a company called Confinity (a combination of the words *confidence* and *infinity*), whose initial objective was to facilitate the transfer of money between two parties via virtual accounts.

The service started operating in 1999, just months before the dot-com crash in the United States (between March 2000 and September 2002, the NASDAQ lost nearly 80% of its value, after having quintupled between 1995 and the start of 2000). But Confinity survived, and a few of its primary executives, like Peter Thiel (Palantir) and Elon Musk (SpaceX, Tesla), continue to work in the ecosystem of innovation. At the end of 2019, the company had more than 300 million active users, transferring money in more than 20 currencies around the world.

Receiving credit card payments has also become simple thanks to another company that did not exist until the first decade of the 2000s. Square—founded in 2009 by Jack Dorsey (Twitter) and Jim McKelvey—enables a smartphone fitted with a small square device (hence the name) to receive card payments.

Analogous to the EdTech sector, the FinTech sector focuses on many aspects of the industry, including two areas that have been given their own names: *InsureTech* (innovations for the insurance industry) and *RegTech* (innovations for monitoring the regulatory aspects of business). FinTech seeks to use technology to reduce complexity, time, and the inefficiencies of antiquated processes. Jobs will be created and destroyed over the course of this transformation as it impacts all aspects of people's relationship with money.

Services such as asset management, real estate investment, payment processing, and capital markets are also being affected by the presence of startups in the sector. But no innovation associated with FinTech has as much broad-reaching potential as *cryptocurrencies*, which we'll discuss next. Whether or not these will prosper is a question that is yet to be answered, but it is possible that some aspects of their implementation—such as blockchain, which we will also discuss—will follow a trajectory of growth.

RAIDERS OF THE LOST COINS

Financial institutions—and banks, in particular—provide a broad range of services, including loans, transfers, payments, and asset management. These services are carefully regulated and overseen around the world, since agricultural and industrial activities and services depend on a healthy financial system. Their primary product is money, which is undergoing significant innovation with the appearance of cryptocurrencies (simply put, a currency that only exists digitally).

Traditionally, financial systems have been centralized and coordinated by the monetary authorities of each country, which also act in an integrated manner on a global scale, with the goal of maintaining the stability, robustness, and security of the system. The so-called monetary base is set by the central banks, which can issue more money, thus providing more liquidity to the system, depending on their interpretation of the economic situation as a whole.

BITCOIN

Bitcoin is the most prominent of a group of digital currencies—money that exists as computer code. It is the best-known implementation of a cryptocurrency and was conceived by a mysterious figure known as Satoshi Nakamoto, who may be one person or several people who worked together for around two years (probably based on the East Coast of the

continued

United States, Central America, or South America) on the model that was launched in 2009. Unlike traditional currencies, there is no control by a central agency: When a new type of cryptocurrency is launched, the rules on the maximum quantity to be issued and the rate at which this value will be reached are preset. Similar to the extraction of precious minerals, which exist in a finite quantity on the planet, cryptocurrencies also have a predetermined supply. And just as miners extract precious stones from the Earth, cryptocurrencies are also found by a community of virtual miners, at a rate also similar to the extraction of gold in the real world.

The extraction of cryptocurrencies is a computationally intensive process, and it is how more monetary units are placed in circulation. There are currently hundreds of types of cryptocurrencies, each one with its own features: ethereum, ripple, litecoin, and monero are just a few examples. Some will survive, while others won't. The valuation of bitcoin (and of cryptocurrencies in general) fluctuates significantly. In December 2012 one bitcoin (BTC) was worth around $13. Three years later, in December 2015, the price was close to $430. At the end of 2016 it was worth something like $960, and over the last month of 2017 it topped $17,000. By mid-2020 around 18.5 million bitcoins were in circulation (worth around $10,000 each), and approximately 2.5 million bitcoins had been mined globally.

But anyone willing to hunt for these coins needs to dedicate time and computational resources to solve complex problems, since creating a safe and fraud-proof environment are key goals of cryptocurrencies. The system for controlling extraction and recording transactions involving cryptocurrencies—including the addition of new currencies onto the market—is implemented via a structure called blockchain—a digital database that contains information that can be used and shared simultaneously in a decentralized, publicly accessible network. The implications of blockchain transcend the world of finance.

As we have seen, one of the most important characteristics of the payment systems today has to do with the centralization of information. During the transfer of funds from a purchaser to a seller, the availability of the funds needs to be verified for the transaction to go through. Traditionally, this is done by consulting a single, centralized database that is typically managed by a bank or the credit card company in question, or by the body that oversees the infrastructure for the selected payment method.

The most important concept introduced by blockchains is a change in the way information is stored. The concept applies to any type of information, but right now the best-known application for this technology is associated with cryptocurrencies: Rather than using a centralized database, exact copies of all available information (or ledgers) are distributed to all parties, using blockchains. The information is not only decentralized, but it is also distributed among all computers that are part of the system. When a new transaction happens, it is validated and inserted into all existing copies.

The name *blockchain* was created based on how the information is stored. Each *block* of information is part of a *chain* established chronologically and protected by cryptographic techniques and sophisticated mathematical functions. In other words, for a new record to be entered into the blockchain, it needs to be validated by all the computers that are part of the network, making fraud or changes to the data that were already stored a complex and costly proposition. In addition to distributing the information, thereby eliminating the risk of a hack on a central point taking down the entire network, blockchain technology also removes the need for an intermediary to serve as a trustworthy party to the transaction—the community itself is responsible for ensuring the system's integrity and robustness.

Digital currencies such as bitcoin use the blockchain structure to enable something that was once impossible: the transfer of funds from one party to another without any type of intermediation. McKinsey & Company estimates that in 2018 the global payments industry saw revenues of almost two trillion dollars in fees. With the increase in the use of new payment methods via cell phone or card (both of which require some type of intermediation), this volume is set to keep growing (in 2009, still according to McKinsey & Company, total revenues were at a trillion dollars). Models such as PayPal, Apple Pay, Samsung Pay, Google Wallet, Stripe, and Square have increased the number of options for transferring money, but they have not significantly modified the basic processing structure for transfers. With a centralized monetary system, even when sensors are used to let a given store know you have purchased a certain item, the way the money leaves your account to get into the store's account is the same as always.

But when cryptocurrencies are used, the process is theoretically more secure, and transferring funds becomes as simple as sending an email. We have become accustomed to working with digital content in many different situations. Sending and receiving letters was a common method of communication just a few decades ago. Movies and TV series, which were once viewed on videotapes, and then on DVDs or Blu-rays, can now be requested on demand via streaming services (where the content is transmitted directly to the device that made the request). Our favorite music is no longer stored in vast collections of CDs, but rather in digital files. And the same goes for our books, sparking heated debates among those who prefer the traditional experience of carrying around several hundred pages, versus the convenience of carrying around a whole library.

THE MONEY MIGRATION

Money has also started its irreversible migration into the digital world: We don't have to go to the bank to pay our bills, make investments, or request a statement—everything is accessible online, via computers, smartphones, and tablets. The staggering growth in the use of credit and debit cards, as well as payments made over cell phones, reduces the handling of paper money that some still insist on carrying around. According to the World Payments Report prepared by French consultancy Capgemini, non-cash transactions reached $539 billion during 2016–2017, and a compounded growth rate of 14% is expected to occur until at least 2022.

The convenience associated with the digitalization of communication, movies, music, and books is now advancing on one of the first and oldest symbols of civilized society. But sending and receiving money is a sensitive matter that requires secure, reliable, and robust technologies.

To exemplify (and simplify) how the process works, imagine you wish to transfer five bitcoins to someone. Using an app on your phone, which contains a unique electronic signature associated with you and this specific transaction, you only need to enter the amount and the account number for the transfer. This instruction will be transmitted to the network, which will validate that you are in fact the sender of the message (using encryption techniques to confirm your identity, as we will see in Chapter 12) and that your balance is sufficient to carry out the transaction (by consulting the shared database with the balance information of all participants). After this, your balance will be reduced by five bitcoins, and your counterpart's balance will increase by five bitcoins. The way this validation occurs is one of the most important characteristics of blockchain: Since the system is completely decentralized and distributed, the responsibility lies with volunteers (called maintainers), who use their computers to autonomously process the transactions done over the network.

Each time a new transaction is performed, all of the maintainers' computers receive this information, meaning that the balances of all participants on the system are available not only in a central database, but in the numerous copies of it that are distributed around the world. These copies need to be synchronized to avoid discrepancies. Imagine, for example, a situation in which a person's balance is equal to 10. This person makes a payment of eight units and, before the update was made, they make another payment of four units. The system cannot allow this to occur, in order to prevent people from spending funds they don't have. The integrity of all copies is, in fact, upheld using a process similar to a vote: As transactions are validated, a complex mathematical function is reprocessed (based on the history of transactions already performed) and compared to the results obtained by the other maintainers. When the network reaches a consensus on the new transaction—only accepting it if it is legitimate—all balances are synchronized, thus avoiding fraud and the improper use of funds.

In the next chapter we will discuss the mechanisms of distribution and security that could potentially make blockchain technology flexible and appealing for use in different applications, as well as the risks of fraud. We'll also talk about the speed and efficiency challenges that this new paradigm imposes on the processing of transactions, as well as the basics of encryption, which is necessary for keeping our digital world secure.

12

ENCRYPTION AND BLOCKCHAIN

TO TRUST OR NOT TO TRUST. WHEN WE THINK ABOUT STORING our confidential information, we normally think of a single, centralized place with restricted access. Imagine the information from your bank statement: It is stored in your bank's systems, and it is accessible with credentials such as passwords and/or biometrics. The idea of concentrating humanity's accumulated knowledge in a few large libraries has been around for thousands of years—one of the most memorable examples is the Great Library of Alexandria, founded in the third century BCE under the leadership of Ptolemy I, successor to Alexander the Great. The library was partially burned during a civil war in 48 BCE and eventually lost its importance over the following four centuries.

But the biggest risk of a highly centralized system is the very fact that there is a clear and obvious target for potential attackers. The origin of the Internet itself came from the need to decentralize communication among military bases, thus preventing an enemy from taking down the entire system with an attack on a single, central point. Although large companies do have redundant systems and protocols to attempt to minimize the effects of a catastrophe (whether natural or not), the potential disruption to clients is still significant.

Centralized systems also need one entity—an institution or a government, for example—to hold power over all the information. Users need to have a relationship of trust with the provider of the information. This means that when we search for a piece of information on our bank's website, we need to believe that the data we find correctly reflects our statement, and that we are in fact communicating with the same institution that we deal with in the physical world. Certifying bodies have been created precisely for this purpose—to attest that a given website does indeed represent who they claim to be. In this model, another trust relationship needs to be established, this time with a certifying authority that validates the website you are visiting. And your device needs to believe that the certificate associated with that website is legitimate and that it correctly specifies that you are dealing with the institution you think you are dealing with.

One way of understanding how this process works is by using the concepts of *public key* and *private key*. Imagine you want to send a digital message to someone (an email, a photo, a video, an audio file—it does not matter), and that you only want the addressee of your message to be able to read it. In addition, you also want the addressee to be certain that the sender of the message was you, and not someone pretending to be you. In other words, you want to guarantee both the confidentiality of the communication and the authenticity of its sender. You also want to communicate with anyone you like, so long as they also have their public and private keys.

THE RSA AND DH ALGORITHMS

The major challenge to the public key encryption model was managing the use of an untrustworthy channel (the Internet) to exchange sensitive information such as users' keys (which are necessary to sign and encrypt messages). The answer to this challenge came in 1978 from two computer scientists—American Ron Rivest and Israeli Adi Shamir—and one mathematician, American Leonard Adleman. Together, they developed and published an algorithm (named after the initials of their last names—RSA) during their studies at MIT. The same solution, developed five years earlier by English mathematician Clifford Cocks during his work at the United Kingdom's Government Communications Headquarters (GCHQ), was classified as confidential and not published until 1997.

continued

Figures 12.1 and 12.2. Ron Rivest, Adi Shamir, and Leonard Adleman, authors of the RSA encryption algorithm. In the photo on the right, English mathematician Clifford Cocks. Source: Wikimedia Commons

The security of the RSA algorithm, which is widely used to encrypt and decrypt digitally transmitted content, is based on a mathematical problem that, up to now, does not have an efficient solution: factorization of very high numbers into prime numbers. The problem consists of finding the prime numbers that, when multiplied together, result in the original number (which is the number used as the key for encrypting and signing messages). One of the most useful features of RSA is that by combining the public and private keys of Users A and B, it is possible to ensure that only User B can read the message from User A, and User B can also be certain that User A is in fact the author of the original message.

Still, despite the efforts of scholars around the world, an algorithm has not been found that could factor a large number into prime numbers within an acceptable time frame. For example, factoring a number containing 232 digits (represented by 768 bits on a computer) was achieved in 2009 after a set of more than 100 computers worked on the problem for two years—and the higher the number of digits, the longer this will take. The critical issue is that it hasn't been mathematically proven that this problem is impossible to solve—in other words, it is not yet possible to determine whether there is, in fact, no efficient way of factoring a number into its prime factors. As we'll see in Chapter 19, the advent of quantum computers—which don't use bits in their on and off states, but rather so-called qubits, which can be both on and off at the same time—is seen by some as a threat to the RSA model, since it is possible that the processing of several types of algorithms may be done much more efficiently using machines of this type.

continued

In the RSA model, each user has two keys: one public key (which guarantees that only the person to whom the message is addressed can read it) and one private key (used to sign and encrypt a message). Although they are mathematically related, the problem of how to efficiently deduce the private key based on the public key has not been (and may never be) solved. However, the encryption and decryption process for all messages using four keys (two public ones and two private ones) is inefficient, so in practice a different solution is used wherein the two parties involved in the communication share a single key. The problem, in this case, becomes how to set this single key in a secure manner.

Once again, the GCHQ team solved the problem first. In 1969, Britons James Ellis (1924–1997), Clifford Cocks, and Malcolm Williamson (1950–2015) developed a solution to the challenge of encrypting a public key, and once again the work was classified as confidential and was only unsealed in 1997 (one month after Ellis's death). The solution, patented in 1977 by Whitfield Diffie, Martin Hellman, and Ralph Merkle, was named the DH key exchange (D for Diffie and H for Hellman; Merkle was Hellman's PhD student).

The DH algorithm is considered to be a *symmetric* type of encryption, since only one key is used to encrypt and decrypt the messages. The RSA solution uses an *asymmetric* type of encryption, with different keys for encrypting and decrypting. Symmetric encryption is faster, but asymmetric encryption is theoretically more secure, especially considering that there is no need for both parties to know the private (and only) key for a given communication.

The prevalent model for secure communications is now a hybrid one, using both symmetric and asymmetric encryption. The primary function of this model, called the public key infrastructure (PKI, which is behind millions of transactions that occur daily over the Internet), is to guarantee that all parties to the communication are who they say they are. This guarantee is done via a certification authority that is recognized by all participants in that communication, and which links keys to users (governments, companies, or banks, for example). Thus, we find ourselves back at the problem of needing a central entity that must be trusted by all participants and which could potentially become a vulnerable point in the process.

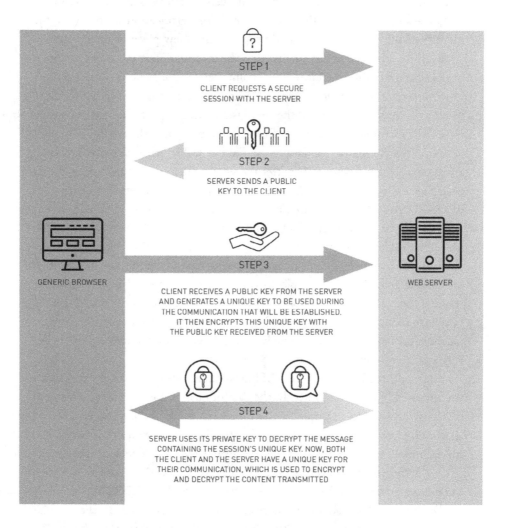

**CLIENTS AND SERVERS ARE CERTIFIED
BY A CENTRAL AUTHORITY THAT VALIDATES
THEIR RESPECTIVE IDENTITIES**

STEP 1

CLIENT REQUESTS A SECURE
SESSION WITH THE SERVER

STEP 2

SERVER SENDS A PUBLIC
KEY TO THE CLIENT

GENERIC BROWSER

STEP 3

CLIENT RECEIVES A PUBLIC KEY FROM THE SERVER
AND GENERATES A UNIQUE KEY TO BE USED DURING
THE COMMUNICATION THAT WILL BE ESTABLISHED.
IT THEN ENCRYPTS THIS UNIQUE KEY WITH
THE PUBLIC KEY RECEIVED FROM THE SERVER

WEB SERVER

STEP 4

SERVER USES ITS PRIVATE KEY TO DECRYPT THE MESSAGE
CONTAINING THE SESSION'S UNIQUE KEY. NOW, BOTH
THE CLIENT AND THE SERVER HAVE A UNIQUE KEY FOR
THEIR COMMUNICATION, WHICH IS USED TO ENCRYPT
AND DECRYPT THE CONTENT TRANSMITTED

Figure 12.3. Example of the use of symmetric encryption in SSL
(secure socket layer) implementation. Source: Author

NO TRUST REQUIRED

Blockchain technology does not require a trust relationship with any central entity. The storage of information is done in a decentralized and distributed fashion, with copies of the data housed in dozens, hundreds, or even thousands of computers. Because of this, there is no central entity that owns the information—and, as a result, there is no need for certification or proof of the identity of the data provider. In addition, when a new piece of information is added, the entire community verifies that the information is valid, and all copies of the blockchain in question—on all computers in the network—are updated. Once a piece of information is inserted into the blockchain, it is very difficult to modify because the integrity of the chain is directly related to the content stored in each block through the use of a function known as a hash.

Hash functions enable a data set of any size to be turned into a sequence of a fixed size. In other words, no matter the size of the information stored, when a hash function is applied to it, the result will be a sequence of letters and numbers of a fixed length. Each block that makes up the structure of a blockchain is connected to the hash of the previous block, and it also contains its own hash, which is based on the chain's entire history. This is an ingenious solution that prevents any piece of information from being changed after it is recorded, because if a piece of data is changed, all subsequent blocks will need to be modified as well (since the result of the hash functions will be modified). To do this, a hacker would need to be able to recalculate and modify all blocks before any new information could be placed in the blockchain—which makes fraud practically impossible, except under very specific conditions, as we will soon see.

Currently, the best-known use of blockchain is for cryptocurrencies, which we discussed in the last chapter. In the case of bitcoin, for example, each stored block corresponds to a transaction performed as though it were an entry in an accounting ledger. This ledger is copied onto all computers on the network, and for each new transaction, all participating machines check to see whether the party spending the bitcoins truly has the balance to make the transaction. When a consensus is reached, the transaction is attached to the blockchain.

For bitcoin's blockchain, the necessary consensus is based on the calculation of the result of the accounting ledger's hash—a process that demands a lot of processing power and that rewards the network's maintainers (or miners) with bitcoins. It is estimated that in early 2017 the energy required to process just one bitcoin transaction was equivalent to the consumption of 4,000 credit card transactions. For this reason, new ways of obtaining consensus among the

BLOCKCHAIN STRUCTURE

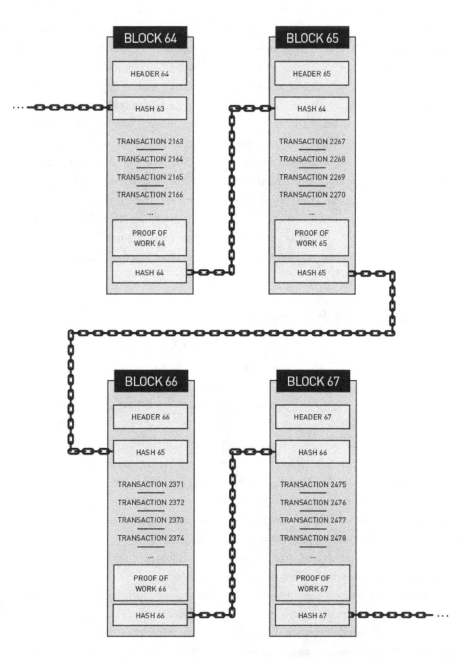

Figure 12.4. Source: Author

parties to the blockchain are being sought, aiming at reducing this technology's energy consumption and the time the network takes to reach a consensus and validate the transaction.

One of blockchain's potential security-related problems—which has already started to be exploited by hackers—is the so-called 51% attack. This happens when an attacker gains control of more than half of the machines making up the blockchain network and causes the majority of the machines to arrive at false results, thus sabotaging the blockchain. In other words, the consensus is no longer trustworthy, and the network has been compromised. This type of attack is more common on relatively small networks, and as blockchains get bigger, the computing power necessary to execute this type of attack becomes prohibitive. Blockchain's technology has characteristics that satisfy a large number of requirements. Blockchains can store any type of information, and they always do it in chronological order. Thanks to the use of encryption, game theory, and efficient communication protocols, it is very difficult—but certainly not impossible—to edit the information once it is stored in one of the blocks in the chain.

BLOCKCHAIN BONANZA

When a technology is a general-purpose one, that is, when it has sufficiently generic characteristics to support multiple types of applications, it is harder to foresee its scope of use. Blockchain fits into this category. In a 2015 report, the World Economic Forum estimated that by 2025 no less than 10% of world GDP may be stored in blockchains.

In late 2016 Deloitte carried out a study with 308 executives from US companies from different sectors, all with annual revenue greater than $500 million: 21% of interviewees stated that they already had systems based on blockchain, and 25% said they planned to implement such systems over the following 12 months. In the health-care sector, 35% of the executives stated that they had plans to produce systems based on blockchain in the short term, while sectors such as consumer goods, technology, media, telecommunications, and financial services indicated an increase in investments directed toward this technology.

The finance sector already has several practical applications of the technology, potentially leading to gains in efficiency and a reduction in costs. The monetary authorities of several countries (including Canada, Singapore, and England) are analyzing the impacts and benefits of adopting the technology for matters such as taxation, international payments, and settlement of operations.

Thanks to blockchain, the possibility of the implementation of micropayments (very low amounts paid for goods or services) is becoming viable for the media, publication, and advertising industries, according to research carried out by Christian Catalini, a professor at the MIT Sloan School of Management. The use of these microtransactions in the world of video games, with the creation of virtual inventories where participants can store their armor, weapons, skills, and accessories, is also likely to benefit from this technology.

Each of the blocks in the blockchain typically represents one transaction, such as the sale or purchase of a certain number of bitcoins, the registration of a patent, a person's credentials, the deed to a property, copyrights to a song or a movie, or a vote for a certain candidate. Whatever the application, the distributed, secure, and chronologically ordered environment enables auditing, trustworthiness, and transparency. It is not surprising, then, that so many industries are interested in applying this technology to their businesses. Transactions performed using blockchain are by nature validated and audited, thus eliminating the costs associated with the verification of the parties, the origins of the goods, payments, and the like.

SMART CONTRACTS

The introduction of blockchain technology has sparked a series of initiatives that make this innovation a powerful tool for the business world. One of these is the Hyperledger project, which was launched in December 2015 by the Linux Foundation and follows the open-code model championed by the same organization. The project is a collaborative effort among a large number of teams who share ideas, suggestions, and models. Participants in this project include Airbus, Baidu, Hitachi, IBM, Intel, JP Morgan, and SAP. All in all, the consortium includes dozens of companies representing different sectors of the economy—aerospace, finance, technology, logistics, and electronics, to name a few. The common goal is improvement to the performance and trustworthiness of blockchains, including services for identification and support for smart contracts.

Smart contracts lay down a series of rules that must be followed for a certain action to be taken, just like a traditional contract—as long as a series of predetermined conditions are met. These conditions may be very broad ranging: the GPS location of a vehicle, indicating that the destination was reached; the receipt of the electronic signatures of all parties to a negotiation; the increase of a price above a certain limit for a certain asset on the financial market; or even confirmation of receipt of a purchase order. As long as a computer is able to interpret the data, the contract will be executed as planned.

According to the World Trade Organization, the impact of improvements to global trade owing to the use of blockchains could be significant, boosting the global volume of goods exchanged, and consequently boosting global GDP. In fact, one of the first concrete results of the Hyperledger project was the result of a partnership between IBM and Danish shipping company Maersk. In the first quarter of 2017, a blockchain-based system was launched to improve the efficiency and security of the exporting and importing process for goods shipped in sea containers, which is how approximately 90% of globally traded items are transported.

Preventing fraud and reducing the processing cost for the documentation associated with each container (which represents around one-fifth of the total shipping cost) are immediate benefits of the system, which was tested on the export of flowers from Kenya to Rotterdam, in the Netherlands. Traditionally, this transaction would generate around 200 communications among the parties involved, but with adequate blockchain support, both the security and the efficiency of the transaction were improved.

In the Kenyan example, workflow is initiated by the exporter, who records a smart contract in the blockchain, which, in turn, initiates the flow of approvals necessary in the three agencies of the Kenyan government that are connected to the process. All parties involved have access to the digital documentation, which is updated with the appropriate signatures (also digital). At the same time, the process for the inspection of the flowers, preparation of the container, and the delivery truck's route are shared electronically with the port of Mombasa, which can take the appropriate measures to receive the container at the appointed time.

Another use for a blockchain structure is the creation of records on the characteristics and origin of products. The English company Everledger, for example, recorded nearly 1.5 million diamonds in a blockchain, storing information such as the number of carats, color, and certification number for each one. By doing this, rather than relying on papers (which can be falsified), all participants in the chain of production, distribution, inspection, and sale can use the records stored in the blockchain. As we have seen, the use of complex mathematical functions ensures that the alteration of the records stored in a blockchain is (at least for now) practically impossible.

The same logic is driving initiatives in the food industry, where firms such as Walmart, Unilever, and Nestlé are seeking a solution that would enable the tracking of each item sold to the consumer, beginning at the very source. This would ensure the quality of the products and immediately pinpoint the source of any health problem caused to any individual consumer or group of consumers (who will probably start storing their medical records in a blockchain in the next few years).

As with any new technology, blockchain faces a set of important questions. Issues such as privacy, anonymity, legal backing, and technical standards are still the subjects of discussion, and all of these issues impact companies' decision-making processes with regard to using blockchain. In the same Deloitte study we referenced earlier (with more than 300 executives from major US companies), 56% of those interviewed believed that the establishment of universal standards for the implementation of blockchains would be a game changer for the adoption of the technology, and 48% reinforced the need for legislation to ensure legal validity in terms of auditing, contracts, and certificates obtained via blockchain. Thinking about matters such as these, a cryptocurrency named ethereum introduced the use of smart contracts in its implementation.

In yet another demonstration of how the technologies introduced by the Fourth Industrial Revolution will undoubtedly be combined to create even more powerful settings for businesses and services, the relationship between blockchain and the IoT has started to yield projects with significant impact and relevance. For example, the measurement of power consumption, and its associated billing, can be done automatically by smart contracts using the data gathered by sensors and stored in a blockchain. In addition, faults or wear of components in the power transmission network could be detected early, sending alerts to the right maintenance teams and thus avoiding power supply problems.

It is important to note that smart contracts are programmed into a blockchain, meaning that any error in the code will lead to unexpected and undesirable results for the parties involved. This happened in at least one high-profile case during the fundraising process for a decentralized autonomous organization in the first half of 2016. The organization had been warned by teams of lawyers about the risks they were exposing their investors to. And, as they predicted, a hacker (or group of hackers) exploited a vulnerability in the code of the smart contract and diverted around $50 million of the more than $150 million that had been raised for the project through a crowdfunding effort.

With the intrinsic characteristics of blockchain, including the difficulty of editing records that have already been inserted and the storage of events in strictly chronological order, the parties involved can, in large measure, trust the information provided. The need for a central authority to validate the data and the need for a complex structure to oversee global supply chains are gradually being eliminated.

A report published by the World Trade Organization indicated that the export of goods from its 164 member countries moved more than $19 trillion in 2019, and the export of services reached almost $6 trillion that same year. As consumer goods are being produced, packaged, distributed, commercialized, and delivered at an unprecedented rate (and several steps in the supply chain are highly suitable for automation), it is possible to imagine that in a not-too-distant future—maybe two or three decades—the only human involved in an e-commerce transaction (or the 3D printing of the desired item) will be the person who placed the order. All the other steps, from production to final delivery, could be the responsibility of machines.

Deloitte and MHI (an international association for the materials, logistics, and supply chain industry) published a study in April 2017 on 900 executives in the areas of manufacturing and supply chain. Responses of 80% of the interviewees estimated that by 2022, the digital supply chain will be the dominant model on the market. Of the remaining 20%, 16% stated that this is already the case. The three technologies that stood out in the study as sources of competitive advantages were predictive analysis, the IoT, and robotics and automation—the topic of our next chapter.

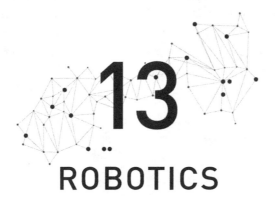

ROBOTICS

IN COUNTLESS CULTURES SINCE ANCIENT TIMES, HUMANS have sought artificial entities that could autonomously carry out repetitive tasks. The histories of Babylon, Greece, India, China, and Europe contain varied and complex examples of this type of entity, both in theory and practice. According to mythology, the Greek god of artisans—Hephaestus—built golden mechanical servants. The fascination and fear we have about beings that are devoid of life, but that carry out tasks at our command, are undoubtedly tied to primitive feelings connected to the act of creation.

One of the first efforts that stemmed from a desire for automation was tracking the passage of time. In the sixteenth century BCE, the Babylonians and the Egyptians developed clocks based on controlled flows of water (some historians believe that the Chinese used a similar technology more than 2,000 years earlier). The water clocks were called *clepsydrae*, which in Greek means "to steal water." Greek inventor Ctesibius (285–222 BCE) refined the technology, incorporated human figurines into the designs, and introduced the concept of pointers to show the time. His work with compressed air and pumps garnered him the title of "father of pneumatics," and the modern-day piano owes a debt to his invention of the pipe organ.

The impact of automata on the future of employment was not lost on the Greek philosopher Aristotle (384–322 BCE), who wrote, "There is only one condition in which we can imagine managers not needing subordinates and masters not needing slaves. This condition would be that each instrument could do its own work, at the word of command or by intelligent anticipation. . . ." In fact, the mathematician Hero of Alexandria (10–70 CE), recognized as one of the greatest experimenters of ancient times, wrote a series of treatises on the work of machines driven by air, steam, or water (*Pneumatica*, influenced by the work of Ctesibius), and machines that opened and closed doors and served food and drinks (*Automata*, from the Greek for "acting of itself").

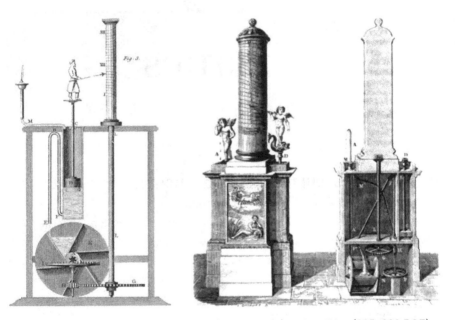

Figures 13.1 and 13.2. The clepsydra (water clock) by Ctesibius (285–222 BCE) from two periods: in an illustration modified from the book by Vitruvius (ca. 70–15 BCE) and as a drawing by French architect Claude Perrault (1613–1688). Sources: Abraham Rees (1819) "Clepsydra" in Cyclopædia; Extrait de "Astronomie populaire" Tome 1 de François ARAGO sur Wikisource (PD-US)

More than a thousand years later in the eleventh century, under the reign of King Bhoja of the Indian Paramara dynasty, the architectural encyclopedic work *Samarangana Sutradhara* devoted a chapter to the development of automata, including bees, birds, fountains, and figurines, both male and female, who refilled oil lamps, danced, and played musical instruments. Inventor Ismail al-Jazari (1136–1206), influenced by both Indian and Chinese culture, wrote *The Book of Knowledge of Ingenious Mechanical Devices*, published in the year of his death, presenting his inventions and showing how to build them. One of his many creations includes what was likely history's first automated musical entertainment system: a boat with four "musicians" operated by hydraulic pumps that produced different rhythms.

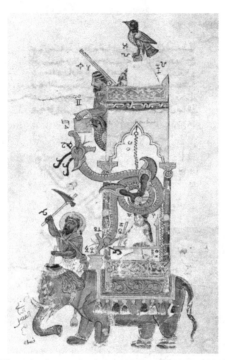

Figure 13.3. The elephant clock by al-Jazari (1136–1206), who wrote: "The elephant represents Indian and African cultures, the two dragons represent ancient Chinese culture, the phoenix represents Persian culture, the water work represents ancient Greek culture, and the turban represents Islamic culture." Source: Folio from Ismail al-Jazari, *The Book of Knowledge of Ingenious Mechanical Devices*, Metropolitan Museum of Art

In 1495, Leonardo da Vinci (1452–1519), Renaissance scientist and artist, designed and possibly built, with pulleys and ropes, a robot knight that could stand, sit, get up, lift its visor, and move its arms and jaw. Its proportions were in line with the Vitruvian Man, which da Vinci had drawn around five years earlier, based on the writings of Vitruvius (70–15 BCE).

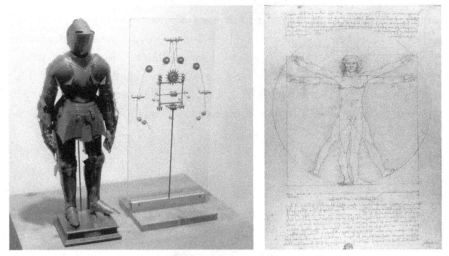

Figures 13.4 and 13.5. Model of the Robot Knight, based on a 1495 design by Leonardo da Vinci, with proportions that were in line with the Vitruvian Man, which da Vinci had drawn around five years earlier, based on the writings of Vitruvius (70–15 BCE). Sources: Photo by Erik Möller, "Leonardo da Vinci. Mensch – Erfinder – Genie" exhibit, Berlin 2005; Photo by Luc Viatour, Gallerie dell'Accademia, 2007

ROBOTIC ENTERTAINMENT

Automata became progressively more sophisticated and were used for entertainment. One of the most famous inventors in this phase of the history of robotics was Frenchman Jacques de Vaucanson (1709–1782), who designed and built three highly complex robotic pieces in 1737. The first, *The Flute Player*, was to scale and moved its fingers to play 12 different songs. (Another famous robot, *The Trumpet Player*, was created by Johann Friedrich Kaufmann, from Germany, in 1810.) Vaucanson next built *The Tambourine Player*, and finished up with his masterpiece, *The Digesting Duck*. The metallic animal could flap its wings, drink water, eat, and simulate the digestive process by excreting material that had been placed in a secret compartment. The book *Karakuri Zui*, an illustrated collection of automata designed in Japan, was published at the end of the eighteenth century, and the future founder of Toshiba, Tanaka Hisashige (1799–1881), was one of the most noted inventors and builders of these machines at the start of his career.

INTERIOR OF VAUCANSON'S AUTOMATIC DUCK.
A, clockwork; *B*, pump; *C*, mill for grinding grain; *F*, intestinal tube;
J, bill; *H*, head; *M*, feet.

Figures 13.6, 13.7, and 13.8. Three examples of automata: *The Digesting Duck*, designed by Jacques de Vaucanson (1709–1782) and exhibited in 1739, in a 1998 replica by artist and restorer Jacques Frédéric Vidoni; *The Trumpet Player*, created in 1810 by Johann Friedrich Kaufmann (1785–1865); and an example from the 1796 book *Karakuri Zui*, a Japanese publication on automata used for different purposes, including serving tea. Sources: Wikimedia Commons, Deutsches Museum, The British Museum

THE MECHANICAL TURK

One of the most famous hoaxes involving an automaton was the *Mechanical Turk*, which operated from 1770 to 1854. The invention by Hungarian Kempelen Farkas (1734–1804) was presented as an automaton that played chess. The giant figure of a Turkish man actually held a human player inside it, but it was convincing to most spectators, who truly believed that the machine itself was moving and playing the game. The invention garnered fame, and some of the Turk's opponents, including Napoleon Bonaparte (1769–1821) and Benjamin Franklin (1706–1790), were unaware they were in fact playing against famous chess players like German Johann Allgaier (1763–1823); William Lewis (1787–1870), from England; or Frenchman Jacques Mouret (1787–1837).

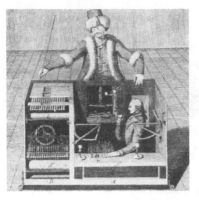

Figure 13.9. The Mechanical Turk, which traveled the world from the late eighteenth to the middle of the nineteenth century. Source: Copper engraving from the book: Freiherr Joseph Friedrich zu Racknitz, *Ueber den Schachspieler des Herrn von Kempelen*, Leipzig und Dresden 1789.

FORCED LABOR

In Czech, the word *robota* means something like "forced labor." The first citation of the English word *robot* dates from 1839, when it was used to describe a labor relationship in Central Europe where a tenant's rent was paid through forced labor. It was not until the twentieth century—more precisely, in 1920—that the word *robot* was used with its current meaning, in a play by Czech author Karel Čapek (1890–1938), following a suggestion from his brother, Josef Čapek (1887–1945).

In the play called *R.U.R.—Rossumovi Univerzální Roboti* (Rossum's Universal Robots), artificial people are manufactured to serve humans, and they end up revolting against the human race—a topic that has repeated itself in science fiction up to current times and that involves profound philosophical and cognitive issues.

R.U.R. was adapted by the BBC in 1938 and became the first science-fiction program to be produced for TV. In the movies, the first robot made its appearance in 1927 in *Metropolis*, by Fritz Lang (1890–1976), based on the book by Thea von Harbou (1888–1954)—the same period in which Japan unveiled the first in a series of famous automata named *Gakutensoku* (meaning "learning from the laws of nature"). Lilliput, probably the first toy robot, was presented for sale in Japan; according to collectors, this happened in the early 1930s, but according to experts, this probably happened after World War II.

Figures 13.10 and 13.11. Scene from the 1921 play *R.U.R.*, which would go on to be translated into 30 languages over the following two years; the play introduced recurring themes in science fiction. On the right, 'Maria' from the movie *Metropolis* (1927), which ushered in a long lineage of robots in film.
Source: Wikimedia Commons

Ironically, the original robots from *R.U.R.* are actually androids because they have a human appearance. In the 1930s, humanoid robots (robots with the same body structure as a person, with a head, arms, and legs, but that don't look like a human) started to become popular. Early examples are Eric and George, created by Englishman William Henry Richards (1868–1948) and Elektro and his dog Sparko, created by Joseph Barnett and produced by Westinghouse. At that time, technology made it possible for robots to talk, sit, get up, and walk.

Figure 13.12. George the Robot and its creator W.H. Richards, in 1930.
Source: Bild Bundesarchiv, German Federal Archives

HERE, THERE, AND EVERYWHERE

Writer Isaac Asimov (1920–1992) created the term *robotics* at the start of the 1940s. One of the most important milestones in its history occurred with the formalization of a new field of study by American mathematician and philosopher Norbert Wiener (1894–1964): *cybernetics*. Cybernetics is the scientific study of how people, animals, and machines control and communicate information. The word comes from the Greek term *kybernétēs*, which means "steersman, governor, pilot, or rudder." The machines we reviewed previously by Ctesibius and al-Jazari may be considered primitive examples of cybernetic systems—even the steam engine itself, by James Watt, had a feedback mechanism to control its speed.

As often happens, one of the first applications of an emerging technology was for military purposes. During the Second World War, Wiener developed a filter at MIT that, when mounted on an anti-aircraft battery, could process signals received via radar to estimate the current position of enemy aircraft. The batteries outfitted with Wiener's filter were more efficient than those without, especially when the target had a generally predictable flight path (such as the V-1 missiles Germany launched against London in the final stages of the conflict).

The common thread among the different fields of cybernetics (engineering, computing, biology, mathematics, psychology, sociology, education, and art) is the processing of feedback signals produced by the very systems being studied and modeled. In his 1950 book, *The Human Use of Human Beings*, Wiener explains the parallels between autonomous systems and society itself, and he highlights ways of preventing people from being compelled to carry out repetitive movements for manual tasks—basically the fundamental objective that drives the development of robots.

But an inventor from Louisville, Kentucky, set the wheels in motion for the robotics revolution in industry. In 1954, George Devol (1912–2011), who worked at the Sperry Gyroscope Company (the very company we discussed in Chapter 2), applied for a patent for what would end up being the first programmable robot. It made its debut in 1961 in a General Motors factory in New Jersey.

The industrial robot, Unimate, was produced by Unimation, a company founded by Joseph Engelberger (1925–2015) and Devol, who had met at a social function in 1956. Ten years later, Engelberger and Unimate appeared on *The Tonight Show*, which was hosted by Johnny Carson (1925–2005) and watched by millions of Americans every night. That same year, researchers from the Stanford Research Institute (affiliated with Stanford University) created a robot named Shakey.

Figure 13.13. George Devol (1912-2011) and his 1954 patent for a machine that could store instructions and move parts. Sources: National Inventors Hall of Fame, Google Patents

Shakey was the first truly autonomous mobile robot able to understand the environment around it and whose programming was almost entirely in LISP. According to scientist Charles Rosen (1917–2002), the project lead, its name was chosen after a month of searching, when one of the members of the team said, "Hey, it shakes like hell and moves around, let's just call it Shakey." In the late 1970s, on the Stanford University campus, a "CAUTION ROBOT VEHICLE" sign warned visitors to keep an eye out for one of Shakey's successors: the bicycle-wheeled Stanford Cart designed to simulate the remote control of a lunar rover from Earth. But its key contribution came in October 1979, when it was able to navigate through a room full of obstacles without human interference. The robot was part of a study by the Stanford Artificial Intelligence Lab (SAIL)—several of whose researchers went on to continue their work on autonomous vehicles at the Robotics Institute at Carnegie Mellon University in Pittsburgh, Pennsylvania, founded that same year.

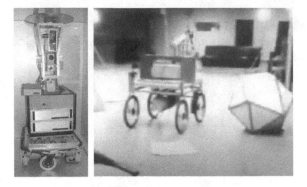

Figures 13.14 and 13.15. Shakey the Robot (1966–1972) and the Stanford Cart (approx. 1970 to 1980). Sources: Wikimedia Commons, Internet Archive

One of the first studies conducted by the Robotics Institute was part of the search for solutions for balancing robotic legs. The simple task of walking and balancing, which humans are able to master at around a year old, is relatively complex for humanoid robots. In 1986, Honda started its development program that eventually led to the creation, nearly 15 years later, of the robot ASIMO (Advanced Step in Innovative Mobility), which became known worldwide. Honda ended production of ASIMO in 2018 to focus on uses of her technology in physical therapy and self-driving vehicles.

Another robot that achieved world fame was Sophia, by Hanson Robotics. Created in 2016, Sophia's camera eyes allow her to see and recognize individuals, and her ability to process speech and conduct conversations brought us one step closer to androids that will be able to play social roles in the future, combining artificial intelligence with image and voice recognition techniques.

Figures 13.16 and 13.17. ASIMO, a robot developed by Honda between 2000 and 2018, and Sophia, created in Hong Kong in 2016 by Hanson Robotics and that is officially a citizen of Saudi Arabia. Source: Shutterstock

AT YOUR SERVICE: A BUTLER IN THE WAREHOUSE

The growing presence of robots in the economy impacts productivity, employability, and reliability in a series of processes. Commercial and industrial robots carry out tasks that are too dangerous or repetitive for humans, and they already work in medicine, industry, logistics, exploration (of land, space, and the ocean), transportation, cleaning, security, entertainment, and agriculture.

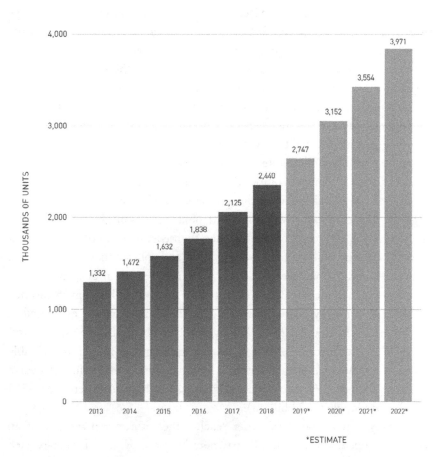

INDUSTRIAL ROBOTS IN THE WORLD'S FACTORIES
2013 TO 2022

*ESTIMATE

Figure 13.18. Source: International Federation of Robotics-World Robotics 2019

Changes in multiple aspects of the technological infrastructure that underpins the world of business are driving initiatives that seek to leverage and reorganize the ways companies operate. For example, the expansion of the IoT has led to the development of communication protocols used exclusively by machines—M2M, or machine to machine—thus enabling the gathering and exchange of information among devices. The multibillion-dollar sensors market enables inventory control systems to determine when it is necessary to generate a new purchase order, and equipment across assembly lines can now act when components demonstrate a behavior that typically precedes a fault.

Modern consumers are accustomed to the convenience of making their purchases online—initially, it was just books (which coexist with their digital counterparts) or CDs (which have undergone a sharp decline because of streaming and downloading services), but now we can purchase almost anything without leaving the comfort of our homes. Food, clothing, toys, school supplies, electronics, cars, or property: All you need is to be connected. This change in consumer behavior has caused a redesign of the entire structure of distribution and logistics: An order can be initiated from anywhere in the world, at any moment, for a specific item from among hundreds of millions of available options just waiting to be selected. It is estimated that Amazon, the global leader in e-commerce, facilitates the purchase of somewhere between four and five hundred million different items on its website.

One of Amazon's major competitive advantages is its distribution facilities. Spread globally—in the United States, Canada, Mexico, Brazil, the United Kingdom, France, Germany, China, and Japan, for example—these centers make it possible for most products to reach their customers in two days or less. The facilities were built by integrating robots and computers into all steps of the process, optimizing aspects such as package size, origin of dispatch, and delivery route, helping the company generate nearly $280 billion in revenue in 2019. The company owns the world's most valuable brand—estimated at $416 billion in the BrandZ 2020 ranking published by British consulting firm Kantar Millward Brown. According to the ranking, the four most valuable brands in the world belong to tech companies, for a total value of over a trillion dollars: after Amazon comes Apple ($352 billion), Microsoft ($327 billion), and Google ($323 billion).

Typically, when a purchase order reaches the information systems of an e-commerce operation, and once its payment has been processed, the next step is to determine which distribution center will dispatch the product to the consumer. This is done by cross-referencing the data on the delivery address and the availability of the item in the warehouses around the country (or even around the world). Some distribution centers can store millions of items with different expiry dates, sizes, weights, and quantities, and one of the main steps in a modern supply chain is related to the efficiency of these facilities.

Consider, for example, the process of picking a specific item, stored on one of the hundreds of shelving units in a distribution center equivalent in size to several city blocks. Not only does the item's location need to be found, but it also needs to be physically grabbed in the most efficient way possible and then packaged and dispatched to its final destination. Among many different methods to accomplish these steps is the use of robots called "butlers," developed by the company GreyOrange, founded in 2011 and headquartered in Singapore. These robots are able to bring the shelving units containing the items to the center's operators in an optimized manner and can potentially fulfill multiple orders simultaneously. Autonomous and available 24 hours a day, the butlers connect themselves to a power outlet when their batteries need to be recharged.

This development was so significant that in 2012 Amazon acquired Kiva Systems for $775 million dollars. Located near Boston, Massachusetts, its name was changed to Amazon Robotics in 2015. This business unit develops robots that help transform Amazon's distribution centers into efficient and precise operations that (using an estimated 200,000 robots spread out around the world) can dispatch millions of orders—almost always with multiple items—on a daily basis.

In 2016 Freightos—a startup founded in 2012 that offers an online marketplace (that is, a setting for negotiations via computer, tablet, or smartphone) for logistics companies—published a study in which executives from the biggest companies of the sector were questioned on which innovations would impact the industry in the most significant manner. Robotics was the technology named most often, in 68% of responses. This was followed by 3D printing (49%), the use of drones and autonomous vehicles (32%), and AR (8%). In fact, currently the most significant investments made by modern distribution centers are related to automation and the seamless integration of robots, sensors, and humans into their operational flow.

MARILYN, KENNEDY, AND THE DRONES

Just like the Internet itself, drones are the result of innovations developed by the military over the course of more than a hundred years. A drone is an unmanned aerial vehicle (UAV) that can fly autonomously or be controlled remotely, but the first use of this idea on the battlefield was not exactly under control. In March of 1848, an uprising against Austrian rule in Venice led to the birth of the Republic of San Marco, a revolutionary state that lasted for 17 months. To crush the insurrection, the Austrian army surrounded the city and launched balloons loaded with explosives, but changes in wind direction ended up sending several balloons the wrong way.

Fifty years later, in 1898, Nikola Tesla (1856–1943), the inventor, engineer, and futurist born in what is now Croatia, presented a small remote-controlled boat during the Electrical Exhibition at Madison Square Garden in New York. This was so extraordinary many considered it to be a magic trick, and not science—corroborating what science-fiction writer Arthur C. Clarke (1917–2008) once said: "Any sufficiently advanced technology is indistinguishable from magic." With his patent, named *Method of and apparatus for controlling mechanism of moving vessels or vehicles,* Tesla created the concept of drones. He wrote in his 1921 book, *My Inventions*: "Teleautomata will be ultimately produced, capable of acting as if possessed of their own intelligence, and their advent will create a revolution."

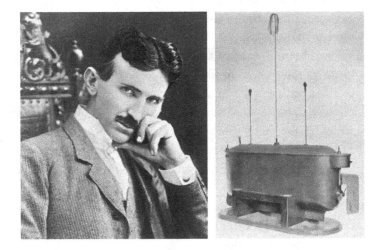

Figures 13.19 and 13.20. Nikola Tesla (1856–1943), one of the most notable inventors in the history of science, and his boat, the first radio-controlled vehicle (1898). Sources: Wikimedia Commons (PD-US); Nikola Tesla Museum, Belgrade (PD-US)

The use of electromagnetic waves to control vehicles from a distance continued to evolve, primarily for military purposes. English engineer and inventor Archibald Low (1888–1956) created a device that paved the way for the development of television and is considered by many to be one of the most important figures in the world of drones. He established the basic principles of remote control to pilot aircraft, and he was responsible for the first tests on the technology in 1917. That same year, the Hewitt-Sperry Automatic Airplane (yes, that's the same Sperry we mentioned in Chapter 2) flew almost 50 km (30 mi) with no pilot and dropped its cargo 3 km (2 mi) from its target.

British actor Reginald Denny (1891–1967), who served in the Royal Air Force during World War I (1914–1918), became interested in radio-controlled airplanes and decided to open a model airplane shop in 1935. During World War II (1939–1945), after signing a contract with the US Army, Denny's company and his partners manufactured around 15,000 drones for the training of operators of anti-aircraft batteries. It was in one of these plants that an army photographer named David Conover (chosen by army captain and future US president Ronald Reagan) saw a young operator named Norma Jeane Mortenson working on the assembly of drones. This was the start of her career as a model, and she later changed her name to Marilyn Monroe (1926–1962).

Figure 13.21. Norma Jeane Mortenson (who later changed her name to Marilyn Monroe) working on the drone assembly line for the US Army. Sources: *Yank, the Army Weekly*, US Army photographer David Conover

The use of drones as weapons of attack by the US Army during this period typically involved a pilot, due to the state of remote-control techniques at the time. The pilot would take off in the aircraft, and once he reached the correct altitude and direction would parachute down into allied territory. Joseph Kennedy Jr. (1915–1944), the older brother of future US president John F. Kennedy (1917–1963), was a drone pilot who ended up dying in a premature detonation. The targets of his attack were German scientists working on their V-2 rocket project; many of these scientists were brought to the United States after World War II to work in the anti-ballistic missile program.

Significant advances in missile development considerably reduced the resources and interest invested in drones for a while, but then technological improvements and the miniaturization of circuits, combined with the need to fight terrorism (urban and otherwise), elevated drones once again to a more prominent position. Drones currently benefit from patents registered in 1940 by Edward Sorensen that make it possible for the operator (or pilot) to have precise information on what is happening with the UAV at every moment, even without visual contact. Their use in the modern military arsenal was expedited in the 1980s by entrepreneur Abraham Karem, who founded the company Leading Systems in his garage in Los Angeles. Using simple materials, he managed to demonstrate the capacity and durability of unmanned vehicles for reconnaissance missions. The research agency at the US Department of Defense (DARPA) financed the work of this startup (which eventually went bankrupt), and this led to the design of modern drones that carry out missions around the world every day.

Drones are following a similar path to ARPANET, the precursor to the current Internet: a technology developed for military purposes that finds its way into society. After a long history of development motivated by war, interest in these unmanned aircraft continues to increase. Consumers wishing to pilot a drone for recreational purposes have dozens of different models at their disposal, with prices ranging from twenty to a few thousand dollars. Many of them have onboard cameras, with different resolutions and ranges depending on their objectives. Obtaining high-definition photos and video is now commonplace, and there are recreational drones with lenses that can comfortably focus on objects more than 1 km (0.6 mi) away. Military drones, even when flying at an altitude of 10 km (33,000 ft), are able to accurately read a car's license plate.

The popularization of drones has occurred so quickly that the FAA (Federal Aviation Administration), the body that regulates civil aviation in the United States, has developed extensive legislation on the subject. With the growing number of drones flying across the skies and collisions and crashes, impacts against buildings or posts, and flights in restricted areas, the FAA now requires a license in many cases. According to the FAA, in September

2020 there were about 191,000 remote pilots certified in the United States and roughly 1.7 million registered drones: 1.2 million recreational and half a million commercial.

PEACE AND WAR: MACHINES HAVE NO FREE WILL

Because of their versatility, drones are now being used by power, insurance, construction, agriculture, research, retail, media, and logistics companies. Their tasks are varied and include the inspection of structures, power lines, and gas pipelines; quality control for roofing; monitoring of the movement of vehicles and people; and delivery of goods and medications. According to Drone Industry Insights, this market will grow from $22.5 billion in 2020 to approximately $43 billion in 2025, at a compounded annual growth rate of almost 14%.

New uses and applications are emerging on a regular basis, frequently through the combination of different components, which have been made affordable thanks to advances in manufacturing processes. And society is set to continue to experience this convergence of technologies over the coming decades—a recurring theme in the Deep Tech Revolution. One example is the use of drones in the supply chain using radio-frequency identification (RFID) technology. This technique works by using the responses that special tags (called RFID tags) can give when an electronic reader is activated to scan the information stored in them. This makes the tags behave as passive elements that don't need batteries, and that only respond when specific radio waves reach them. One of the projects underway at MIT uses a fleet of drones that fly autonomously around a warehouse, stimulating the RFID tags with radio waves and cataloging the location of each item stored, with a margin of error of less than 20 cm (8 in).

The integration of artificial intelligence mechanisms and weapons offers the possibility of truly autonomous weapons (autonomous weapons systems or lethal autonomous weapons). An armed drone equipped with facial recognition software could be programmed to kill a certain person or group of people and then to self-destruct, making it practically impossible to determine its source. Machines do not have their own free will; they always follow the instructions of their programmers.

These arms present significant risks, even when used only for defense purposes (a tenuous line for sure), and they evoke images of the killer robots that science-fiction authors have been writing about for decades. In 2015, at the

24th International Joint Conference on Artificial Intelligence, a letter advocating that this type of weapon be abolished was signed by theoretical physicist Stephen Hawking (1942–2018), entrepreneur Elon Musk, and neuroscientist Demis Hassabis (one of the founders of DeepMind, which was acquired by Google in 2014), among others.

The discussion is still ongoing, but there are historical examples that speak to the benefits of the involvement of humans such as Vasili Arkhipov (1926–1998) in life-and-death decisions. In April of 1962, a group of Cuban exiles sponsored by the US Central Intelligence Agency failed in their attempt to invade the Bay of Pigs in Cuba. To prevent a future invasion, the Cuban government asked the Soviet Union to install nuclear missiles on the island.

After obtaining unequivocal proof that these missiles were in fact being installed, the United States mounted a naval blockade to prevent more missiles from getting to the island and demanded the removal of those that had already been installed, which were just 150 km (90 mi) from Florida. In October of 1962, the world watched as tensions between the United States and the Soviet Union mounted and reached their peak.

On October 27, when a Soviet B-59 submarine was located in nearby international waters, a crew from the US Navy dropped depth charges near the vessel to force it to surface. With no contact from Moscow for several days and unable to use the radio, the submarine's captain, Valentin Savitsky, was convinced that the Third World War had begun, and he wanted to launch a nuclear torpedo against the Americans.

But the decision to launch a nuclear weapon from the B-59 needed to be unanimous among the three officials: Captain Savitsky, political officer Ivan Maslennikov, and second-in-command Vasili Arkhipov, who was only 39 at the time. He was the only one to dissent and recommend that the submarine surface in order to contact Moscow. Despite evidence that pointed to war, Arkhipov remained firm and actually saved the world from a nuclear conflict.

14

NANOTECHNOLOGY

IN MAY 1998, A SCIENTIFIC ARTICLE TITLED "RECOMBINANT
Growth" was published in the *Quarterly Journal of Economics*, an important publication of the Department of Economics at Harvard University. Its author, Martin Weitzman, who earned a PhD in economics from MIT in 1967, suggested that the limits to the growth of an economy may not be determined by the capacity for generation of new ideas, but rather by the capacity to transform these ideas into practical and usable projects. Weitzman created a mathematical model in which elements that already existed in the economy—industries, equipment, automobiles, and laboratories, for example—were expanded and combined with innovations, thus creating a new pattern of growth and evolution based on different combinations of all of these components.

According to the Greek philosopher Heraclitus (c. 535–475 BCE), "everything flows"—in other words, the only constant is change, which is at the very heart of the history of civilization.

Our age is no different; in fact, innovations and advances should continue to accelerate due to the combination (and recombination) of new and existing technologies. The individual manipulation of elements (such as sensors, memories, processors, data, or networks) and the creation of systems, programs, equipment, or devices have paved the way for new business segments: autonomous vehicles, 3D printers, robots and drones, smart cities, agrotechnology, and more. But what would happen if, rather than just developing new technologies and combining them among themselves, we decided to also manipulate and combine the basic building blocks of matter—atoms—in a manner similar to what is being done in the field of biotechnology with DNA?

THE NEW MARBLES

The idea of manipulating atoms for the creation of new materials and machines was articulated for the first time by American physicist Richard Feynman (1918–1988) in one of his famous talks. On December 29, 1959, his presentation at the annual meeting of the American Physical Society was transcribed with the title "There's Plenty of Room at the Bottom." In it, he asked the audience "Why cannot we write the entire 24 volumes of the Encyclopedia Britannica on the head of a pin?" And he proposed two challenges.

The first consisted of the construction of an electric engine that would fit within a cube measuring 0.4 mm (0.016 in) on each side—something that was done less than a year later by engineer William McLellan (1924–2011). The second, directly related to Feynman's initial question, involved reducing the size of a text by 25,000 times—but the equipment and techniques available to turn this vision into reality would still take time to materialize. It wasn't until 1980 that it became possible to see atoms and their bonds in detail, through the use of a scanning tunneling microscope. This achievement landed the equipment's inventors, German physicist Gerd Binnig and Swiss physicist Heinrich Rohrer (1933–2013), both from the IBM Research Lab in Zurich, the 1986 Nobel Prize in Physics.

And it took until 1985 for Tom Newman, a PhD candidate at the Department of Electrical Engineering at Stanford University, to overcome the second challenge posed by Feynman.

The term *nanotechnology* was used for the first time in 1974 by a professor at the Tokyo University of Science named Norio Taniguchi (1912–1999), but it started to become more well-known in 1986, when engineer Kim Eric Drexler published *Engines of Creation: The Coming Era of Nanotechnology*, which was revised and expanded in a new edition in 2007. Drexler earned his PhD at MIT, where he was a student of Marvin Minsky (1927–2016), one of the most important names in the field of artificial intelligence.

Figure 14.1. The first page of Charles Dickens's 1859 work *A Tale of Two Cities* reduced 25,000 times using an electron beam on a square of plastic with sides measuring 0.2 mm (0.008 in). Source: American Physical Society, Image: Tom Newman

Nanotechnology deals with objects on an atomic scale, measured in nanometers, or one-billionth (10^{-9}) of a meter. A typical piece of paper is more than 100,000 nanometers thick. In an article published in *National Geographic* in June 2006, reporter Jennifer Kahn illustrated the size difference between a nanometer and a meter: She said it would be the equivalent of comparing a marble to planet Earth.

THE NANO MARKET

As expected, the emergence of a new science with such significant potential gives rise to concerns. After all, what are the possible consequences to our health and our ecosystem of developing new materials or modifying existing ones? The Royal Society, founded in London in 1660 with the objective of promoting scientific knowledge, published a study on the matter in July 2004, called *Nanoscience and nanotechnologies: opportunities and uncertainties,* which laid out not only the potential benefits of the new technology, but also the need for regulation and control in terms of human exposure to nanoparticles.

A NANOMETRIC PERSPECTIVE

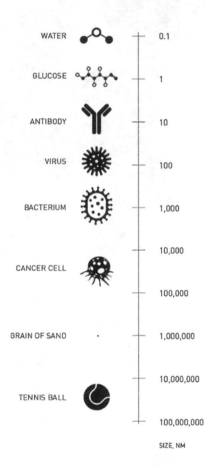

	SIZE, NM
WATER	0.1
GLUCOSE	1
ANTIBODY	10
VIRUS	100
BACTERIUM	1,000
	10,000
CANCER CELL	100,000
GRAIN OF SAND	1,000,000
	10,000,000
TENNIS BALL	100,000,000

1 NANOMETER = ONE-BILLIONTH OF A METER

OUR FINGERNAILS GROW ONE NANOMETER PER SECOND

A STRAND OF HAIR IS APPROX. 80,000-NM THICK

Figure 14.2. Sources: Wikipedia, others

Woodrow Wilson (1856–1924), the 28th president of the United States, led the nation from 1913 to 1921. He had strong ties to the academic world; he held a PhD in political science from Johns Hopkins University and taught at several post-secondary institutions. He served as president of Princeton University between 1902 and 1910, and as of the publication of this book, he is the only US president with a PhD. One of the world's most important think tanks was given his name in 1968: the Woodrow Wilson International Center for Scholars. Think tanks are study centers for the discussion of ideas and solutions; they involve different sectors of society and seek to address high-impact challenges. (It should be noted that, since at least 2015, Princeton students have demanded a name change for the Woodrow Wilson School of Public and International Affairs due to his support of racial segregation. In June 2020, Princeton's Office of Communications announced the new name was going to be The Princeton School of Public and International Affairs.)

In 2005 a partnership between the Wilson Center and a nonprofit established by the Pew family, which founded the petrochemical company Sunoco (now part of Energy Transfer Partners) in the late 1800s, created the Project on Emerging Nanotechnologies. This partnership led to an important piece of reference material for the field: a directory of products containing nanomaterials, targeted to end consumers (the Nanotechnology Consumer Products Inventory). Since this is a new area, whose effects on health and on the environment still need to be monitored and understood, initiatives of this nature are very important. Clearly reflecting what is expected from governments in terms of basic research and development of new technologies, in 2000 then-US president Bill Clinton stated that "Some of these [nanotechnology] research goals will take 20 or more years to achieve, but that is why . . . there is such a critical role for the federal government." Between 2001 and 2020, the US-based Nanotechnology Initiative received almost $29 billion in public investments.

Governments from different countries are allocating resources to the subject because of its complexity and strategic importance. In an article published in July 2016, UNESCO noted that the list of patents and articles in the area of nanotechnology is dominated by the United States, Japan, South Korea, Germany, Switzerland, and France. Three years later, the top five countries in number of nanotech published patents at the US Patent and Trademark Office and the European Patent Office were the United States, South Korea, Japan, China, and Taiwan. Two countries that don't often appear on the lists of the most innovative places, but that are showing considerable growth in research and development in the sector, are Malaysia and Iran; the latter country, like the

United States, also maintains an open catalog of nanotechnology-based products (the Nanotechnology Product Database).

Regardless of how countries decide to invest in *nano*, or their respective talents for research, this technology's field of application includes many sectors, such as energy, manufacturing, agriculture, metallurgy, medicine, engineering, and molecular biology.

THE REAL GLITTER OF GOLD

In the world of nanotechnology—in which typical dimensions are literally a million times smaller than an ant, and tens of thousands of times smaller than the diameter of a strand of hair or the thickness of a piece of paper—the way materials react could be quite different from what we are used to, with their own unique physical, chemical, and biological properties in gas, liquid, or solid state. When we raise the temperature of water above 100°C (212°F), we observe its transformation from a liquid state to a gaseous state. If we do the opposite process and reduce the temperature to below 0°C (32°F), we again modify its initial state, this time from liquid to solid. Our knowledge of the properties of elements and compounds enables us to predict and use the behavior of groups of atoms and molecules to our benefit: conducting electricity, emitting different electromagnetic fields, interacting with greater reactivity with other elements, or exhibiting new ways of reflecting light.

The study of the properties of different types of materials, combined with the capacity for manipulation and grouping of particles on a nanometric scale, opens up new possibilities for multiple lines of business. Gold, for example, is a metal whose nanometric particles absorb light (unlike the behavior observed under the conditions we're used to)—and that light is transformed into heat in a quantity large enough to eliminate undesired cells from the body.

In the field of manufacturing, some of the most often used nano structures are carbon nanotubes, which are formed by a hexagonal arrangement of just one layer of atoms of this element. In addition to being transparent, this material is a good conductor of heat and electricity and, as per the definition from *Oxford Learner's Dictionaries*, it is the thinnest and strongest material known to science. Originally observed in the 1960s, its importance was recognized with the award of the 2010 Nobel Prize in Physics to Russian-born Andre Geim and Konstantin Novoselov for their "groundbreaking experiments regarding the two-dimensional material graphene." Nanotubes are now present in products ranging from bicycles to surf boards and ships to turbines. Research on how to combine carbon nanotubes with plastics seeks to replace steel with a lighter and stronger compound, and several groups are already working with these materials in the field of bioengineering, aiming at making advances in procedures for bone reconstruction and recovery.

The textile industry may have possible uses for nanofibers, which can be combined with regular fibers to give new functionalities to the finished product, such as superhydrophobic clothing—fabrics that repel water and remain dry in virtually any situation. Another example is clothing that curbs unpleasant odors through the application of copper-coated silica nanoparticles.

Nanotechnology, with its capacity to impact multiple industries, is already present in integrated circuits and prosthetics, inks and packaging, fertilizers and fabrics—with new applications under study. Gold, silver, titanium dioxide, and silicon nanoparticles are found in cosmetics, dentistry products, and oil-based lubricants. As of September 2020, the Nanotechnology Products Database available at the StatNano website (a body supported by the Iran Nanotechnology Innovation Council) covered almost 9,000 globally available nanotechnology-based products. New careers and specializations are set to multiply in a labor market that will be undergoing substantial changes in the coming decades.

But risks to human health and the environment are critical issues because, as we have seen, elements and compounds at the nanometric scale behave differently than they do in their macro settings. Because of their size, nanomaterials may be absorbed quickly by our bodies (something that may not always be desirable), with impacts on digestive and circulatory processes. However, the benefits of the technology seem to strongly outweigh its risks, especially in medicine where innovations are being applied in sensors, imaging scans, and targeted releases of medicines into patients with serious illnesses.

With the use of nanoparticles (based on lipids or polymers), it is becoming possible to specifically target the cells that are sick, thus reducing or even eliminating the adverse effects of the medication. Thanks to their size, nanoparticles are able to penetrate the walls of blood vessels, releasing the medication directly inside the tumor. Scientists are seeking to refine this and other methods that promise to significantly advance the quality of existing treatments.

SAFE WATER

According to a 2017 WHO report, nearly 850 million people around the world do not have access to safe drinking water, and at least two billion people consume contaminated water, which is responsible for half a million deaths annually. The report states that in less than 10 years, around half the world's population will be living in water-stressed regions.

Nanomaterials are being explored for the creation of cheaper and more efficient filtration systems: Carbon-based nanomembranes already exist for the desalination and purification of water, and nanometric sensors can detect the presence of bacteria or toxins. Compounds such as titanium dioxide (which can be found in certain solar filters, for example) have demonstrated their capacity to neutralize bacteria such as *Escherichia coli*, which is found in the intestines of many types of animals.

This is actually what technology should be all about—solving critical problems, improving the lives of millions, expanding our collective minds, and building, one step at a time, a staircase toward a better future. The actual origin of the word comes from the Greek *tekhnologia*—a combination of the words skill and collection. Thus, technology is basically a concrete, palpable manifestation of a skill: not only here, on Earth, but anywhere in the universe. And another major challenge that nanotechnology is addressing is the development of solutions for space exploration—the subject of our next chapter.

15

AIRPLANES, ROCKETS, AND SATELLITES

IN 1873, MARK TWAIN (1835–1910) WROTE *THE GILDED AGE: A Tale of Today,* in partnership with his friend Charles Warner (1829–1900). The title refers to the application of thin layers of gold onto a less valuable object (wood or porcelain, for example) as a sort of metaphor for society's problems in the latter third of the nineteenth century. Things looked good on the surface, but structurally, significant problems needed to be addressed. The post–Civil War era in the United States saw the emergence of families with vast fortunes and the expansion of the social gap between the richest and the poorest. Names such as Rockefeller, Mellon, Carnegie, Morgan, and Vanderbilt heavily influenced the course society would take through their investments in railways, metalworking, industry, and banks.

They also laid the groundwork for the American industrial complex and for a culture of corporate philanthropy, which survives today. Bill Gates (the founder of Microsoft) and Warren Buffett (investor and CEO of Berkshire Hathaway) are examples of modern-day philanthropists. Together they have donated more than $80 billion to different causes related to health, education, poverty, and sanitation. A new group of entrepreneurs is ready to influence the future of space exploration—a segment that was, until recently, restricted to the governments of military powers such as the former Soviet Union and the United States.

THE NEW SPACE RACE

Since there is as yet no definitive legislation established, the definition of where space actually begins varies, depending on the institution or the country. NASA considers flights over 80 km (50 mi) above sea level to be spaceflights, while the *Fédération Aéronautique Internationale* (the World Air Sports Federation, founded in 1905 and headquartered in Switzerland) makes this distinction at 100 km (62 mi) above sea level. But the best-known attempt to scientifically establish the definition is attributed to Hungarian-American aerospace engineer Theodore von Kármán (1881–1963).

The so-called Kármán line was calculated in 1956 at 83.8 km (52.1 mi) and defined by Kármán as the maximum altitude at which an aircraft can still rely on the atmosphere to support its flight. In fact, starting at this altitude, Earth's atmosphere is much less significant than the force of gravity. To go into orbit, a vehicle needs to overcome Earth's gravity and maintain a certain velocity.

WHERE DOES SPACE BEGIN?

Legislation applicable to aircraft is separate from that of spacecraft, and as with international waters, which do not belong to any single country, it is crucial to establish where each nation's airspace ends. According to Thomas Gangale, executive director of the OPS-Alaska (Oceanic, Polar, Space) research network, during the NASA Space Shuttle Era (1981–2011), the spacecraft passed just 34 km (21 mi) outside of Cuban airspace on its return to Earth. According to him, space law should govern any space flight, irrespective of what countries are overflown. With the increase in commercial flights and new types of space vehicles under development, it is likely that on the route to space or during the return, a spacecraft will violate the airspace of another country.

continued

Violations of space limits could lead to political disputes with unforeseeable consequences, as happened during the Cold War. When the United States and the former Soviet Union were competing for military and ideological dominance around the world, they ended up in a conflict that reached beyond the limits of Earth itself. Describing the situation, former Canadian prime minister and winner of the 1957 Nobel Peace Prize Lester Pearson (1897–1972) pointed to the replacement of a *balance of power* with a *balance of terror*: a direct confrontation between the two superpowers that would result in mutual annihilation—a war with no winners.

Just as the origins of the Internet lie in military-led projects, the beginnings of space exploration were motivated by the quest for global geopolitical hegemony—which has also been reflected in regional conflicts, such as the wars in Korea, Afghanistan, and Vietnam; the division of Germany into East and West; and the Cuban missile crisis.

The Soviet Union made the initial leaps beyond the Kármán line with the launch of the first satellite, Sputnik 1, in 1957, followed by the first mammal to orbit Earth (Laika the dog, on board Sputnik 2, also in 1957), and, finally, the first human to be sent into orbit, a feat that immortalized Yuri Gagarin (1934–1968) in April of 1961. The following month, Alan Shepard (1923–1998) became the first American to be launched into space, and in September of 1962, then-US president John F. Kennedy made his famous speech dubbed "We Choose to Go to the Moon" in Houston, Texas, and set the end of the 1960s as the deadline to accomplish this aspiration. And indeed, in July 1969, Neil Armstrong (1930–2012), the commander of NASA's Apollo 11 mission, became the first person to walk on a celestial body other than the Earth.

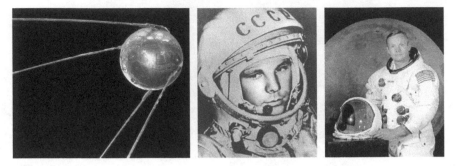

Figures 15.1, 15.2, and 15.3. Left: Sputnik 1, the first artificial satellite to orbit Earth, in 1957. Center: Yuri Gagarin (1934–1968), Soviet cosmonaut and the first human to be sent into Earth's orbit. Gagarin died in a crash during a training flight, at just 34. Right: American astronaut Neil Armstrong (1930–2012), the first human to walk on the Moon. Sources: NASA (PD-USGov); Smithsonian's National Air and Space Museum, NASA Photo ID: S69-31741—Program: Apolo XI (PD-USGov)

Six years after the United States won the race to the Moon, the United States-Soviet rivalry and competition eventually led to collaboration and cooperation: In July 1975, the command modules for the Apollo (United States) and Soyuz (Soviet Union) spacecraft docked, and three Americans and two Soviets worked together for nearly two days, even giving interviews. The handshake between Tom Stafford and Aleksey Leonov marked what many deemed was the end of the Space Race, as well as the foundation for the development of programs like the Mir space station (1986–2001) and the International Space Station (ISS).

This new friendship formed alongside several factors. Technically, the United States had the best space exploration technology, while the Soviet Union focused its research on the resistance and endurance of humans and machines over long periods away from Earth. Politically, an agreement in 1972 signed between the two superpowers (the result of the Strategic Arms Limitation Talks—SALT) and the 1975 Helsinki Accords (involving both Germanies) contributed to the easing of global tensions. And financially, the growing research and development costs to sustain the space programs favored the use of greater efficiency and cooperation between the players.

The collapse of the communist regime, symbolized by the fall of the Berlin Wall in 1989, and the subsequent dissolution of the Soviet Union in 1991 led to new challenges for the world economic and political order, inevitably impacting the course of collaborations and projects. The end of the American Space Shuttle program in 2011, the dependence on the Russian Soyuz system (operated by state-run corporation Roscosmos) for the transport of passengers to the International Space Station (ISS), the lack of definition of Russia's role in the future of the ISS, and the entry of China into the group of countries with the capability to send people into orbit (starting with Yang Liwei in 2003) have set the stage for the next chapter of the human adventure in space.

The first act of the Space Race was characterized by competition between the two governments, and the second one by collaboration and a search for synergies. But we are now in the middle of the third act, and new players are coming onto the stage, seemingly seeking the leading role in this significant business opportunity.

AN OTHER-WORLDLY BUSINESS

The private sector's interest in space exploration is as old as the Space Race itself. Just eight months after cosmonaut Yuri Gagarin became the first human to orbit Earth in April 1961, a satellite designed and built by a group of American amateur radio operators was placed into orbit. The OSCAR I (Orbiting Satellite Carrying Amateur Radio) was launched from a US Air Force rocket, and over 22 days—from December 12, 1961, to January 3, 1962—it transmitted the greeting "Hi" in Morse code.

And not surprisingly, the start of commercial space exploration came from the telecommunications sector. In July 1962, the Telstar 1, a communications satellite (that was active for only seven months) enabled the live transmission of images of the United States to Europe. In late August of the same year, US president John F. Kennedy signed a bill for a law that would regulate the use of space in the telecommunications market.

But it's one thing to develop satellites; placing them in orbit is quite another. The force of gravity, which literally keeps our feet on the ground, has to be overcome—and this is only possible at an escape velocity of at least 11.2 km/s (more than 40,000 km/h, or 25,000 mph). The energy required is significant, and constructing rockets with that capability is generally associated with the governments of a handful of nations.

In 1975, Germany's Lutz Kayser (1939–2017) started the first private initiative for the development and manufacture of engines capable of reaching escape velocity. His company, called OTRAG (Orbital Transport und Raketen), was the legitimate predecessor of SpaceX, founded by Elon Musk (also the founder of Tesla) and of Blue Origin, founded by Jeff Bezos (also the founder of Amazon). OTRAG ceased its operations in 1981, primarily owing to geopolitical pressures, especially from France and the Soviet Union, who opposed the development of Germany's rocket launching capabilities.

Figures 15.4, 15.5, and 15.6. Oscar (1961), Telstar (1962), OTRAG (1977). Sources: Smithsonian's National Air and Space Museum (photo by Eric Long); NASA (PD-USGov); space.skyrocket.de

A very unusual and rather specialized service may be obtained from US-based company Space Services, founded in 1981. Through its subsidiary Celestis (1994), a funeral in space is available. By making use of extra cargo space on launches by other companies (such as Lockheed Martin, Northrop Grumman, and SpaceX), individuals able to pay may have a deceased person's ashes scattered into the cosmos. But Space Services' biggest achievement was probably its 1982 launch of the first privately financed rocket to reach space and stay there for more than 10 minutes.

Another interesting launch involves The Conestoga 1—a name that references the wagons used by the settlers of the American West in the 1800s—that was built out of spare parts and launched from a ranch in Texas. The engines from the second stage of a Minuteman rocket acquired from NASA were used as propulsion. Because NASA was not legally able to sell them (since they were used to launch nuclear bombs), a creative leasing agreement allowed Space Services to pay the full cost in the event the engines were not returned in perfect working condition. If the launch failed, the rocket would explode; if everything went correctly (as turned out to be the case), the rocket would return to its final resting place, in the Gulf of Mexico. That was how a company with seven employees (among them, former astronaut Donald "Deke" Slayton [1924–1993]) and 57 investors who allocated $15 million to the project (adjusted for inflation) were able to pave the way for private space travel.

A year after Space Services was founded, Orbital Sciences Corporation emerged (part of the Northrop Grumman group since 2018), ensuring its place in the history of space exploration with the first launch vehicle entirely designed, developed, implemented, and operated by a private company ever to travel into space. Their Pegasus rocket is launched from an airplane at an altitude of 12 km (7.58 mi) and can carry loads of 440 kg (1,000 lb) into low Earth orbit, that is, up to 2,000 km (1,250 mi) of altitude.

Figures 15.7 and 15.8. Conestoga 1 (1982) and Pegasus (1990). Sources: Space Vector Corporation/Eric Grabow; NASA/Photo ID: GPN-2003-00045 (PD-USGov)

CAPITALIZING ON SPACE

To understand the origins of the next round of innovations in space exploration, we need to go back in time around a hundred years. In 1919, French hotelier Raymond Orteig (1870–1939) offered a prize of $25,000 (currently about $360,000) to the first person who could make the first non-stop flight from New York to Paris (or vice versa). The exploit would be good for his business (his hotel was in New York, where he lived) and would allow him to stay close to the aviation industry, which fascinated him. Charles Lindbergh (1902–1974), a US airmail pilot, won the prize in 1927 after completing the 5,800-km (3,600-mi) journey in around 33 hours on board the famous single-engine *Spirit of St. Louis*.

Technically, Lindbergh's was not actually the first crossing of the Atlantic by air. John Alcock (1892–1919) and Arthur Brown (1886–1948), from England and Scotland, respectively, left St. John's, Newfoundland, and landed in Clifden, Ireland, after completing the more than 3,000-km (1,900-mi) journey in approximately 16 hours in 1919. Then-United Kingdom Secretary of State for Air

Winston Churchill (1874–1965) presented the prize of £10,000 (today around £500,000) offered by the *Daily Mail* to whoever could accomplish the feat.

Nearly 80 years after it was created, the Orteig Prize served as inspiration for the XPRIZE. This competition, sponsored by the nonprofit foundation begun in 1995 by Greek-American entrepreneur Peter Diamandis, aims at facilitating radical advances for the benefit of humanity. Over its more than 20 years of existence, the themes of the competitions have included the development of cars with exceptional fuel efficiency, the creation of new technologies to deal with oil spills in the ocean, and the use of sensors to monitor health.

But the first, and likely the most famous, XPRIZE contest was about space exploration. The challenge, announced in 1996, was to build a private space-craft capable of transporting three people and of flying twice in two weeks. SpaceShipOne—the name of the vehicle built by Mojave Aerospace Ventures (by Microsoft cofounder and billionaire Paul Allen [1953–2018] and aerospace engineer Burt Rutan) passed the Kármán line on September 29 and October 4, 2004. Mojave Aerospace Ventures won the $10-million prize (although SpaceShipOne consumed around $25 million). The funds to pay out this prize came from the Iranian-American Ansari family, thus ensuring that the process would run without any government subsidy.

Figure 15.9. SpaceShipOne. Source: Flight 15P, Wikimedia Commons, photo by D. Ramey Logan

In 2007, it was Google's turn to issue a challenge, with the Google Lunar $20 million XPRIZE, to successfully launch, land, and operate a vehicle on the Moon's surface. The prize expired in 2018, however, with no team achieving the goal.

The XPRIZE has garnered a lot of media attention and inspired entrepreneurs, businesspeople, and inventors and also enabled the start of a new age in space exploration, financed by billionaires such as Allen, Richard Branson (who obtained the license for the SpaceShipOne technology and created Virgin Galactic), Jeff Bezos (Blue Origin), and Elon Musk (SpaceX). As we'll see a little further on, space exploration by private companies significantly impacts many different business segments here on Earth—from multipurpose satellites, mining asteroids, spacecraft for cargo transport, and defense systems to space tourism and commercial flights.

FIRST-MOVER (DIS)ADVANTAGE

The expression *first-mover advantage* is occasionally used to explain the success of some companies that position themselves in a certain market before anyone else. With their ability to provide unique technology or to acquire strategic resources before prices go up, these companies theoretically pull out front first, potentially winning consumers who will think twice before switching to the competitors that will inevitably follow.

But this advantage does not always translate to long-term success, and newcomer competitors often end up benefiting from the mistakes and the victories of those who waded into unexplored waters before them. Let's use Space Services as an example: In 1982, it became the first private company to launch a rocket into space. Three years later, the US government awarded it the first license as a service provider for launching commercial rockets—and in March of 1989, Starfire became the first spacecraft to fly with this type of authorization. But less than 18 months later, Space Services' investors halted financing activities for the company, and it became a secondary player in a market that was starting to gain significant traction: launching satellites for telecommunications, TV, navigation, and scientific research.

Currently, one of the most critical aspects for the progress of the space exploration business depends on the cost associated with sending cargo into low Earth orbit, which is defined as orbital altitudes of less than 2,000 km (1,200 mi). The vast majority of objects in orbit are located in this region, including the ISS and the Hubble telescope. In the early 1980s, the cost per kilogram of cargo on the space shuttle was greater than $80,000. By the mid-1990s, it was less than $27,000. In 2009, Falcon 1, by SpaceX, brought this price down to less than $10,000, and in 2017, Falcon 9 broke the $2,000 barrier. In just four decades, the cost of putting objects in orbit has fallen by over 97%.

The opportunity for capitalization on the space industry is immense. This market is currently estimated at around $350 million—a figure that is set to rise to somewhere between $1 trillion (according to Morgan Stanley) and over $3 trillion (according to Bank of America Merrill Lynch) by 2040. A large portion of this revenue will likely come from the launch of satellites which, among other things, will potentially broaden access to the Internet around the world, provide high-resolution images of Earth's surface, and inform our devices of their precise positioning.

Other companies with relevant roles (at least for now) in the new space race include Orbital ATK (founded in 1982, a subsidiary of the Northrop Grumman group since 2018), United Launch Alliance (or ULA, founded in 2006 as a joint venture between Lockheed Martin and Boeing), and Sierra Nevada (which in 2008 acquired SpaceDev, the developer of components for Paul Allen's SpaceShipOne, winner of the first XPRIZE).

FASTEN YOUR SEATBELTS

Space exploration by private companies also impacts commercial aviation. June of 2019 marked the 100-year anniversary of the first transatlantic flight by Alcock and Brown. Nearly 20 years after this accomplishment, the first commercial transatlantic flights were made in the Boeing B-314, also known as flying boats (since they took off from and landed on water). The flight from Southampton, England, to New York took around 30 hours, with three stops.

The Douglas DC-4 (produced in 1921 by the Douglas Aircraft Company, which merged with McDonnell Aircraft in 1967, forming McDonnell Douglas, which in turn became part of Boeing in 1997) was the first airplane to regularly carry out transatlantic flights. In 14 hours, you could fly from New York to Bournemouth, England, with two stops. The next innovation—the

LARGER LOADS, LOWER COSTS

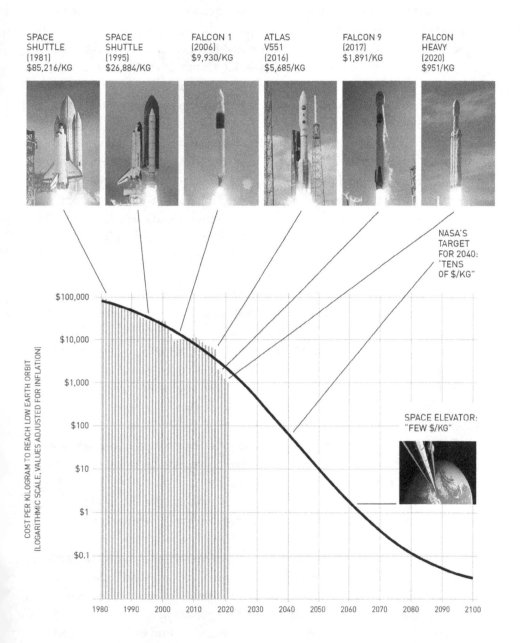

Figure 15.10. Sources: FutureTimeline.net; Wikipedia; NASA

introduction of the jet engine in the early 1950s—reduced flight times significantly, thus decisively consolidating commercial passenger aviation: It was possible to fly from New York to Paris in just under eight hours.

Since then, both regional and international flight times have not shortened significantly—except for the Concorde (a joint venture of the French state-run corporation Aérospatiale and the British Aircraft Corporation), which operated between 1976 and 2003. This aircraft reached supersonic speeds and typically cut the flight times of other types of planes in half.

Unfortunately, the environmental cost of transporting people from one place to another—whether by car, truck, train, boat, or airplane—is high. According to the US Environmental Protection Agency (EPA), the transportation sector accounted for almost 30% of greenhouse gas emissions in the United States in 2017, and more than 90% of fuel is still petroleum based (including gas and diesel). We'll talk more about the greenhouse effect and its relationship to technology in Chapter 16.

Different methods with promise in reducing the environmental impact of the aeronautical industry include the use of lighter materials, use of new types of fuels, optimization of consumption, installation of solar panels in major airports (whose daily power consumption is equivalent to a city of 100,000 inhabitants), and modifications to aerodynamics. The installation of winglets—small structures installed on the ends of the wings—has led to an approximately 6% reduction in fuel consumption per flight, and studies on new layouts for aircraft (including blended wing body) promise a reduction in weight of almost 15% and in fuel consumption of almost 30%.

But there are even bolder proposals under discussion. Over the next few decades, long-haul flights or flights with high passenger traffic may see their routes diverted through space—potentially making it possible to travel any distance in less than three hours. There are certainly economic justifications for this, considering the number of passengers and the costs associated with air travel, as well as the substantial reduction in flight times. But before this happens, the safety of flights crossing the Kármán line still needs to improve, with a drastic reduction in the hitches encountered in all stages of the operation—after all, for the time being, operating and flying airplanes is a process much more common and less prone to errors than operating and flying spacecraft.

Another idea with a science-fiction feel is space tourism. In 2001, the world watched as the first tourist reached the ISS: Dennis Tito, from the United States, took a ride in the Soyuz capsule for a price of $20 million. Another six tourists partook in this experience over the course of the first decade of this century, but since 2009, there have been no more spots for civilians on flights to the ISS. Virgin Galactic claims it has already sold more than 600 tickets (at between $200,000 and $250,000 each) for future flights, where passengers will experience zero gravity and a spectacular view of our planet, before returning to Earth. At some point, Blue Origin will also be set to offer flights like this.

While the popularization of passenger transport in space—whether as a form of innovation in the aviation market, or as a new type of tourism—still needs a few more years to mature, the same cannot be said of the economic revolution caused by *smallsats* (small satellites).

CELESTIAL ACCOMPLICES

The Renaissance pulled humanity out of the ignorance and superstition of the Middle Ages toward a world grounded in the arts and sciences, and its history is forever linked to the Medicis, the family who dominated the political and economic scene in the Italian city-state of Florence for nearly three hundred years (between 1434 and 1737). Their wealth originated from the fabric trade and grew with the Banco Medici (1397–1494), one of the most respected financial institutions of its day. By sponsoring artists and scientists, the family provided a fertile setting for the development of masters such as architect Filippo Brunelleschi (1377–1446), the versatile Leonardo da Vinci (1452–1519), and painter Sandro Botticelli (1445–1510). Cosimo II de' Medici (1590–1621) was the Grand Duke of Tuscany from 1609 up to his death. One of his tutors was none other than Italian astronomer and physicist Galileo Galilei (1564–1642), the pioneer of both science grounded in detailed and thorough observations and of the scientific method (defined as "systematic observation, measurement, and experiment, and the formulation, testing, and modification of hypotheses").

In January of 1610, Galileo used his recently built telescope to observe four "stars" whose relative positions to Jupiter were inexplicably changing. Within a week, he figured out that the stars were actually the moons of Jupiter in their orbit.

In Latin, the word *satellitem* can be translated as "companion, accomplice, or attendant," and in fourteenth-century French, the term *satellite* referred to a "follower or attendant of a superior person." To honor Cosimo II and his three brothers, Galileo named the four moons *Sidera Medicæa* (the Medician Stars). In 1614, German astronomer Simon Marius (1573–1625) published his book *Mundus Iovialis*, in which he named the satellites Io, Europa, Ganymede, and Callisto after mythical figures who "pleased the god Jupiter."

Galileo's observations of the moons in orbit around Jupiter was devastating to the common beliefs of the times: How could there be celestial bodies that did not orbit Earth—which was then considered to be the center of the universe? His discovery forever changed the geocentric vision of the solar system, where Earth played the central role, and proved the heliocentric vision, in which the sun is at the center. Even though Greek astronomer Aristarchus of Samos had suggested this same vision between 300 and 200 BCE, the world was forced to wait nearly 2,000 years for the publication of *De revolutionibus orbium coelestium* (*On the Revolutions of the Celestial Spheres*) in 1543, by Polish astronomer Nicolaus Copernicus (1473–1543), who formulated a model where the sun, and not Earth, was placed at the center of the universe.

After German astronomer Johannes Kepler (1571–1630) established the laws of planetary motion, Sir Isaac Newton (1643–1727) in 1687 wrote one of the most important books in the history of science: *Philosophiæ Naturalis Principia Mathematica* (*Mathematical Principles of Natural Philosophy*). In it, among other things, Newton introduces the law of universal gravitation, which states that every particle is attracted to every other particle in the universe by a force that is directly proportional to the product of the particles' masses and inversely proportional to the square of the distance between their centers.

Figures 15.11, 15.12, and 15.13. Galileo Galilei (1564–1642),
Nicolaus Copernicus (1473–1543), Johannes Kepler (1571–1630).
Sources: Galileo, oil painting by Justus Sustermans, c. 1637, Uffizi Gallery,
Florence; Copernicus portrait from Town Hall in Toruń, Poland, 1580;
Portrait of Kepler, copy of a lost original from 1610 in the Benedictine
monastery in Kremsmünster, Austria

LAUNCHED OBJECTS

The first use of the word satellite to describe a machine orbiting Earth may have been by French writer Jules Verne (1828–1905) in his 1879 book *The Begum's Fortune* (where Begum is used to designate an aristocrat of South Asian origin). Curiously, this was the author's first publication that took a pessimistic view of technology. Fiction became fact almost 80 years later, when the Soviet Union put Sputnik 1 into Earth's orbit.

The United Nations' Office for Outer Space Affairs has maintained a registry of objects launched from Earth since 1962: As of late March 2020, 9,456 objects had been launched into space. Of these, 5,774 individual satellites were still orbiting Earth, 2,666 of which were still active.

The applications for satellites (civil or military) are varied: communications, navigation, meteorology, the mapping of space, or the mapping of Earth. Depending on their objective, they use different energy systems, altitude control technologies, antennae for transmitting and receiving data, and information-gathering devices.

According to the nonprofit organization Union of Concerned Scientists (founded in Cambridge, Massachusetts, in 1969 and boasting more than 200,000 associates who advocate for science), communications satellites account for around 40% of the active satellites currently in orbit. In fact, the telecommunications market—which is currently responsible for global revenues of over one trillion dollars—has been significantly impacted by space exploration. The United States launched the first communications satellite into space in December 1958. SCORE (Signal Communications by Orbiting Relay Equipment) used short waves to transmit the voice of President Dwight Eisenhower (1890–1969) to Earth. It is interesting to note that this was one of the first projects carried out by ARPA (now DARPA, an agency of the US Department of Defense).

Until the advent of satellites, Earth's curvature had been an insurmountable obstacle for telecommunications devices based on high-frequency radio waves because they needed to have a line of sight between the transmitter and the receiver. Satellites relay the signal around the curve of the planet using antennae, which make communication between distant geographical points possible.

The biggest turning point for this transition was likely in July 1962, with the launch of the Telstar 1. It was operated by telecommunications company AT&T (American Telephone and Telegraph Company, which was founded in 1885 by Alexander Graham Bell [1847–1922], holder of the patent for the first telephone). Telstar 1 was the first satellite to retransmit images (photographs and videos) and sounds (for telephony) and the first to transmit a television program from one side of the Atlantic to the other.

PICK YOUR ORBIT

Satellites generally travel in one of four types of orbits:

1. Elliptical (less than 1,000 km [600 mi] or more than 40,000 km [25,000 mi] above Earth)
2. Geostationary (approximately 36,000 km [22,000 mi] above Earth, a special subset of geosynchronous satellites orbits directly above the equator)
3. Medium (between 8,000 km [5,000 mi] and 24,000 km [15,000 mi] above Earth)
4. Low (between 250 km [150 mi] and 2,000 km [12,000 mi] above Earth); more than 60% of active satellites are in low orbits, and around 30% are in an orbital period matching Earth's rotation (geosynchronous), meaning they take exactly one day to orbit Earth

The idea of communications satellites in geostationary orbits was discussed in 1945 by science-fiction writer Arthur C. Clarke (1917–2008)—based on publications by Austro-Hungarian-born physicist Hermann Oberth (1894–1989) and Slovene engineer Herman PotoÐnik (1892–1929)—with the primary objective being to facilitate communications on Earth. This works because antennae installed at stations on Earth do not need to move to follow the satellite, since both the antenna and the satellite move synchronously with the rotation of the planet.

NEVER LOST

A constellation of 24 satellites in low orbit literally take us where we need to go every day. They tell us exactly where we are and how to get to our destinations. Few technologies have become as popular in so little time as GPS. Its initial designs date from the 1960s, but the system did not become fully operational until 1995, and with the expansion of smartphones and consumer navigation apps, billions of people don't go anywhere without it.

Each satellite in the constellation covers a relatively small area, which changes as it travels at high speed around Earth—that is how they collectively provide continuous coverage. Because they are closer to Earth's surface, the time a signal takes to travel from the satellite in low orbit to Earth's surface (and vice versa) is less than what it would take to make the same return trip via geostationary satellites, for example. This technology that allows us to navigate by land, sea, or air has also made it economically feasible to track farm vehicles, cars, telephones, skateboards, bicycles, ships, drones, airplanes, or anything else that needs to be located and/or guided.

WATCHING THE EARTH

In addition to telecommunications and navigation satellites, nearly one-third of the satellites actively orbiting Earth were developed to observe its surface. Their growth has been notable: According to a Radiant Earth Foundation report based on data from the Union for Concerned Scientists, between 2003 and 2017 almost 600 Earth observation satellites were launched, compared to fewer than 30 in the previous decade. Two key factors are driving the increase of satellites in orbit: the greater number of options for sending them to space and, even more importantly, the development of private companies (of which many are startups) that design and launch small satellites (weighing less than 500 kg, or 1,100 lb).

Some Earth observation satellites take frequent, high-resolution pictures of, for example, shopping center parking lots, which enables analysts to estimate the purchasing health of the economy. Others take photos of forests to track deforestation activities; still others are instrumental in weather forecasting, providing significant benefits to agriculture and transportation. The fact is that satellites have become yet another business sector made possible thanks to technological advances in other fields, including manufacturing, electronics, signal processing, propulsion systems, optics, and aerospace engineering. Take, for example, the period between early 2019 and early 2020: There was a 55% increase in the number of communication satellites, a 24.5% increase in Earth observation satellites, and a 40% increase in technology development/demonstration satellites. In its 2019 annual report, the Satellite Industry Association, formed in 1995 by several American companies, estimated that around 80% of the revenues from the space industry—estimated at $360 billion in 2018—can be directly linked to satellites (think telecommunications, security, science, and observation) and to the manufacturing of the satellites themselves and their auxiliary equipment. Of the $126 billion generated, almost $100 billion was from services for the television sector, and there are sizable expectations for growth of revenues associated with global access to the Internet.

DEEP IMPACT

In 1961, American astronomer Frank Drake derived an equation whose objective was to estimate how many potentially intelligent civilizations exist in the Milky Way. As one of the founders of the SETI project (the search for extraterrestrial intelligence), Drake worked with one of the biggest science communicators of this century, astronomer Carl Sagan (1934–1996), on the development of the first message sent out from Earth. That information was displayed in a gold-anodized aluminum plaque affixed to the outside of the Pioneer 10 and Pioneer 11 probes (launched in 1972 and 1973, respectively), with information on our location in the universe.

Drake's equation has served as a basis for successive discussions and refinements. It takes seven factors into account, including how many planets on average can support life in each solar system, how many life forms effectively evolve and become intelligent, and what percentage of these civilizations develops technologies that could be detected. All of the parameters in the equation are difficult to estimate, and the results obtained typically range from 1,000 to 100,000,000 possible intelligent neighbors in our galaxy. (Astronomers estimate there are more than one hundred billion galaxies, based on data obtained by the Hubble Space Telescope, which was launched in 1990.)

There are convincing reasons to search for new habitable planets: One is the occurrence of extinction (or extinction-level) events. In 1982, paleontologists David Raup (1933–2015) and Jack Sepkoski (1948–1999) identified the five largest of these to take place on Earth; the first occurred 450 million years ago, and the most recent one occurred 66 million years ago. In general, there are multiple causes for extinction events: Volcanic eruptions, cooling of the atmosphere, asteroid impacts, and global warming are some of them.

A scenario in which we're struck by an asteroid has been the topic of films such as *Armageddon* and *Deep Impact*, both from 1998. Whether or not this is a coincidence, 1998 was the year when the United States Congress directed that NASA should be able to detect any asteroid more than 1 km (0.62 mi) in diameter passing less than 200,000 km (125,000 mi) from the sun. The Center for Near-Earth Objects Studies was assigned this task, and in the past 20 years it has detected around 20,000 objects, most of them asteroids. It is believed that the extinction of the dinosaurs was due to an impact from an asteroid or comet around 15 km (9 mi) in diameter that cast off enough ash, dust, and debris into the atmosphere to block solar radiation from Earth for decades.

The inquisitive nature of humanity has forever made us explore the world around us—on the land, on the sea, in the air, and, more recently, in space. Sadly, our search for more efficient ways of traveling through space and eventual settlements of colonies on other planets may end up becoming less a matter of curiosity, and more a matter of survival. The economic impact of the unprecedented exploitation of Earth's natural resources, as well as the opportunities for critical innovation, will be discussed in the next chapter, about energy.

16

ENERGY

ALONG WITH HEALTH CARE AND FOOD, ENERGY IS ONE OF MOD-
ern civilization's biggest needs. With our growing dependence on technology, the
efficient use of energy and the development of equipment to produce, store, and
distribute it are absolutely critical. The global energy industry—which includes
companies that work with fossil fuel, electrical, nuclear, and renewable energy—
saw, in 2018, $1.85 trillion in investments according to the IEA: $771 billion in the
power sector, $726 billion in oil and gas supply, $240 billion in energy efficiency,
$80 billion in coal supply, and $25 billion in renewables for transport and heat.

In its "Statistical Review of World Energy 2020," British Petroleum published its estimate on the percentages of different energy sources making up total
world consumption. In 2019, oil accounted for one-third, followed by coal (27%)
and natural gas (24%)—that is, these three sources represent almost 85% of our
energy use. The rest is distributed among hydroelectric (6.4%), nuclear (4.3%), and
renewable sources such as solar, wind, and geothermal plus biofuels (5%).

Technically, hydroelectric energy can be considered a renewable source. But,
according to a 2016 study by researchers at Washington State University, two
aspects are of concern to environmentalists: the impact of the reservoirs that
feed the turbines and the high levels of methane emissions from these reservoirs
(likely due to decomposing vegetation and the accumulation of nutrients).

ENERGY SOURCES
2019

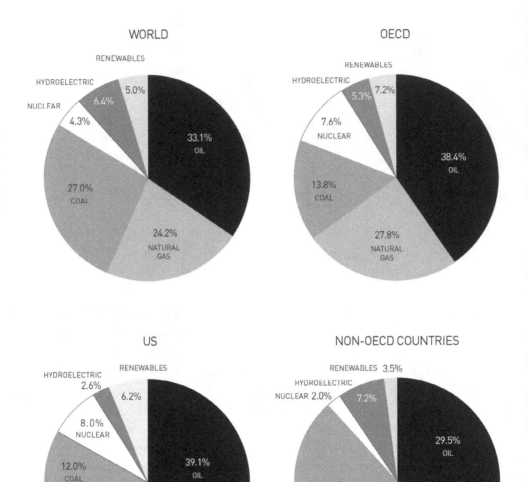

Figure 16.1. Source: BP Statistical Review of World Energy 2020

The United States Department of Energy, via its Energy Information Administration (EIA), estimates that global energy consumption will grow by around 30% between 2015 and 2040, in large part due to the continuous economic growth of countries like India and China. To meet this energy demand, the agency estimates that the percentage of renewable sources in the global energy network will grow, increasing by around 2.3% per year over the same period. Nuclear energy will also grow, at 1.5% per year. Nonetheless, fossil fuels are still forecast to account for 75% of our energy consumption in 2040.

Labor shifting from the industrial sector to the services sector also affects nations' energy patterns. The term *energy intensity* measures the relationship between energy and GDP, and according to a report published by consulting firm McKinsey in late 2016, this measure of efficiency continues to improve. In 2015, for example, the production of one unit of global GDP required nearly one-third less energy than it did in 1990. The same report states that by 2050, only half of the energy used in 2013 will be required to produce one unit of global GDP.

The IEA, formed in 1974 in the wake of the oil crisis, had 30 member countries, eight association countries, and two accession countries by April 2020 seeking to work both on energy-related and environmental issues (which are becoming increasingly difficult to separate). In one of its reports, the body noted that in 2017 only 19% of global energy was used as electricity, while the other 81% was used for transport and heating systems. Despite the improving energy efficiency of electronic devices, it is clear there will be a significant increase in demand for electricity over the coming decades, especially in the case of developing countries. For air conditioning alone, demand is set to more than quadruple, reaching almost 2.5 billion units by 2040.

Although the different players in the energy industry do not agree on when global oil demand will reach its peak, they do agree it is just a matter of time. The most aggressive forecasts place this date at 2023, but most experts estimate it will happen sometime in the 2030s. Fossil fuels will gradually decrease in importance as the number of electric vehicles on the roads increases, as the cost of energy from renewable sources becomes more competitive, and as internal combustion engines become more efficient, thus reducing fuel consumption.

Countries such as Germany, the United Kingdom, France, Norway, and India have already set timelines to end the sale of fossil fuel–burning cars over the next two decades. But it is important to note that there are other sectors—such as aviation and petrochemicals (plastic, cosmetics, fertilizers)—that are likely to inhibit an abrupt drop in oil prices because of the sheer size of their demand.

CHARGED UP

Global support for energy systems that are efficient, clean, and safe for the environment coexists with a growing demand for energy to charge batteries for phones, laptops, and tablets, and soon for our cars and household generators. A common scene at airports, offices, homes, and restaurants is anxious tech users looking for outlets to charge their devices (until wireless recharging becomes more commonplace). Grand View Research estimated the global battery market size at about $108 billion in 2019 and projects a 14% compound annual growth until 2027.

GLOBAL CONSUMPTION
BY TYPE OF ENERGY SOURCE—1990 TO 2040
(IN MILLIONS OF TONNES OF OIL EQUIVALENT)

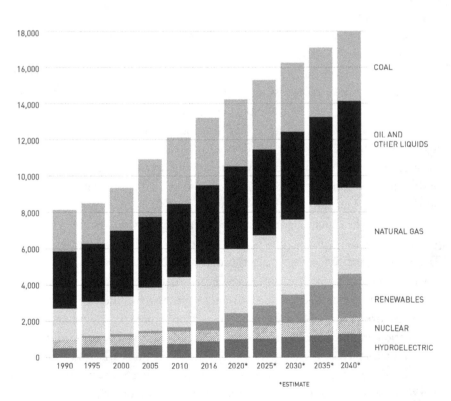

Figure 16.2. Sources: British Petroleum, Energy Information Administration

Producing energy without impacting the environment is critical to the survival of civilization. Even with the anticipated significant increase in electric vehicles, which will favorably impact our ecosystem, batteries will (obviously) need power to be recharged. Currently, the vast majority of batteries in portable electronics and transportation are lithium ion. An ion is basically an atom with an electric charge and, depending on the application, the battery may contain elements such as cobalt, iron, phosphorus, manganese, and nickel. Initially proposed in the 1970s by English chemist Stanley Whittingham, lithium batteries were researched and analyzed for decades (their first commercial launch was in 1991); eventually, they became virtually omnipresent in the equipment that defines modern society. (Whittingham won the Nobel Prize in Chemistry in 2019 for his work, together with John Goodenough and Akira Yoshino.)

The best solution for recycling batteries has not yet been determined. Unfortunately, it is currently around five times cheaper to extract lithium from nature than to effectively recycle an old battery. Unless the ways in which we currently make batteries change, we need to fix this situation if we are to avoid a shortage of elements such as cobalt, nickel, and lithium itself. Applications for lithium-ion batteries are not limited to just small-scale equipment thanks to their longevity, strength, large number of possible charge cycles, and (except in the case of manufacturing flaws) their safety. They are used in electric vehicles and in projects to complement and support the supply system for traditional energy, like the battery bank installed by Elon Musk's Tesla in South Australia.

A POWER BET

In 2017, South Australia planned to increase its grid stability to avoid outages during bad weather, so the government requested proposals for a grid-connected battery solution that would work like a back-up energy system. On March 10, 2017, Musk committed to installing a battery bank with a capacity of 100 MW (enough to fulfill the demand of around 30,000 houses for an hour) in a maximum of a hundred days. If he missed the deadline, he promised to do the project at no cost to the Australian government. But 63 days later, the installation was complete. The batteries were connected to a wind farm owned by the French company Neoen (around 200 km [125 mi] from Adelaide); Neoen sends energy to the system as needed, helping to prevent blackouts and leveling out the supply.

According to the EIA, in 2013 the world consumed about 567 EJ (one exajoule is roughly the equivalent of the energy contained in 174 million barrels of oil). The solar energy absorbed by the atmosphere, oceans, and continents every 90 minutes is around 660 EJ. In other words, Earth receives more energy in an hour and a half than its people consume in a year. The technological challenges to transforming this amazing potential into electricity lie both in the solar panels themselves, which are currently somewhere between 15% and 20% efficient in converting solar radiation into electricity (there are cases where rates of 40% are achieved), and in the batteries' energy storage capacity.

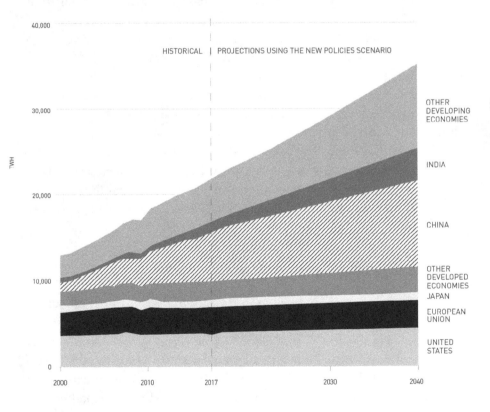

Figure 16.3. The estimate for global electricity demand uses the New Policies Scenario by the International Energy Agency. This scenario includes the policies and targets announced by governments for reducing carbon dioxide emissions. Source: International Energy Agency

PATHWAYS FOR ENERGY

Streets, neighborhoods, cities, states, and countries depend on efficient and robust data communication networks to run public utilities such as water, lighting, sewer, power, transportation, and telecommunications. These networks also serve as the backbone for businesses in the financial, logistics, manufacturing, and services sectors. Our destiny is inextricably linked to the success of this complex patchwork of operating systems, communication protocols, microprocessors, sensors, storage devices, databases, wires, cables, batteries, generators, and transmission lines.

Ever since the electrification of the world, which basically defined the Second Industrial Revolution (the late nineteenth to the early twentieth century), we just assume that we will have electrical energy at our disposal anywhere we go. We have come to inescapably rely on energy generation in a world that is grounded in vast data communication networks for our equipment, appliances, machines, and cities. This brings us to data centers—creations of a new world order governed by information.

Computers that are part of a data center may serve a single company or they may serve multiple end users. They are the physical representation of the concept of cloud computing: gigantic collections of machines and storage drives working 24 hours a day, seven days a week, 365 days a year that literally keep the Internet running. Everything—or almost everything you do—on your personal computer, your smartphone, your tablet, or your smart TV uses machines housed in data centers to respond to your requests. These machines, the storage drives where your remote data is kept, the network equipment that connects all these devices to the Internet, and the cooling equipment required to dissipate the generated heat—all consume electricity.

Research published in 2016 by the Lawrence Berkeley National Laboratory ("United States Data Center Energy Usage Report," authored by Arman Shehabi and others) estimated energy consumed by American data centers at around 70 billion kWh in 2014, giving us some idea of the energy cost that makes our new lifestyles possible. It represented roughly 2% of the country's total electricity consumption, and it included servers, data storage units, network equipment, and infrastructure. With the increase in global data consumption, the need to properly estimate how data centers are expected to grow becomes a key component of a sustainable and energy-efficient society.

According to the same report, data center electricity consumption increased by around only 4% between 2010 and 2014, a big change in relation to the estimated 24% increase between 2005 and 2010 and to the estimated 90% increase between 2000 and 2005. A 2020 article entitled "Recalibrating global data center energy-use estimates" by Eric Masanet (Northwestern University) and others published in *Science* magazine also highlighted the slower growth in data centers' electricity consumption: a relatively modest 6% between 2010 and 2018, when global data centers were estimated to have used 205 TWh (terawatt-hours), or about 1% of global electricity.

A number of factors explain this welcome phenomenon, each one of them addressing key components of electricity use in data centers: servers (responsible for about 43% of the consumption), cooling systems (another 43%), storage drives (11%), and network equipment (3%).

The number of physical servers in use has declined, due in part to the use of virtualization software (which allows one machine to emulate the behavior of several individual servers). Servers are generally allocated to large data centers and are optimized for high usage rates and efficient energy use (although increasingly powerful, servers have required practically the same amount of energy since 2005). The increase in the capacity of storage devices has reduced the number of physical drives necessary.

Keeping data center machines' temperature under control is essential and is something that can be controlled by selecting locations with colder climates or by using artificial intelligence. In 2016, Google started using the DeepMind system to simulate adjustments to the cooling systems in its data centers, and since 2018, it started letting the algorithm effectively control the temperatures.

A reasonable approximation of the global consumption of electrical energy in the second half of the 2010s was 20,000 TWh, and of over 200 countries, the two leaders—China and the United States, respectively in first and second place—accounted for almost half of this value. Information and communications technology represented around 10% of this figure, or 2,000 TWh, and as we have seen, data centers' share of this is 1% of the total, or 200 TWh.

CLICKING CLEAN

One of the most interesting characteristics of technology in general is that, apart from occasional exceptions, the next version—whether it is a simple improvement or a major upgrade—provides improvements upon the previous version while simultaneously bringing problems that were not there before. The replacement of horses with cars for transportation enabled us to get from one place to another faster, but it created significant fossil fuel pollution problems.

Autonomous vehicles could, in theory, significantly reduce overall pollution emissions since it is likely that they (and non-autonomous cars, trucks, and buses, too) will switch to electric. And their shared and coordinated use facilitated by algorithms could make bottlenecks a thing of the past. However a paper published in April 2016 by researchers Zia Wadud (University of Leeds, in the United Kingdom), Don MacKenzie (University of Washington), and Paul Leiby (Oak Ridge National Laboratory, Tennessee) predicts increased travel could cancel out the benefits afforded by automation.

Improved building management technology may help to significantly increase energy efficiency, thus reducing waste and environmentally harmful emissions. Buildings have been responsible for around 60% of the increase in global electricity demand since the end of the twentieth century, so smart heating and cooling systems, along with sensors for monitoring and control, could lead to reduced energy demand.

The fact is that the growth in the number of computers, telephones, sensors, and TVs/monitors is not likely to slow down anytime soon. A 2015 report by Huawei, the Chinese multinational specializing in telecommunications equipment and services, attempted to estimate the impact of this growth. The paper, entitled "On Global Electricity Usage of Communication Technology: Trends to 2030," written by Anders Andrae and Tomas Edler, estimated not only the consumption by data centers, but also the energy cost of the production of communication equipment, of data networks (both wireless and wired), and of the devices used in our day-to-day lives (we can group all of these into information and communications technology, or ICT).

According to Andrae and Edler, if no improvements are made to the efficiency of data networks and devices (which is extremely unlikely, as we have just seen), ITC may account for nearly half of global electricity consumption by 2030. (They expect that this number will end up landing at around 21%, and that the best scenario will be no less than 8%.) Almost half of the world's 50 largest tech companies (based on annual revenue) have set targets for the use of renewable energy as a response to their environmental impact from the high energy consumption of their activities. It is quite possible that this trend will drive investment in and development of renewable energy in the coming years, something that reports like *Clicking Clean*, by environmental organization Greenpeace, tries to monitor closely.

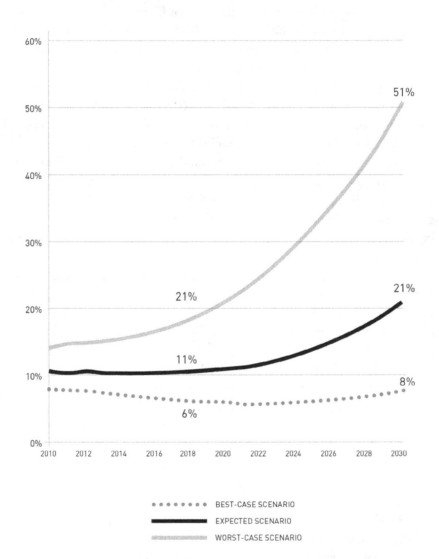

Figure 16.4. Using 2010 as a starting point, when data points were relatively close to each other, researchers have attempted to estimate the percentage of global electricity necessary to run information and communications technology systems. Source: Anders S. G. Andrae and Tomas Edler, "On Global Electricity Usage of Communication Technology: Trends to 2030"

GREENHOUSE BLUES

Governments, companies, and not-for-profits are concerned with the technological challenges associated with the generation, distribution, storage, and efficient use of energy. Regardless of the means, our connected future will continue to demand innovations in this sector. Ever since the First Industrial Revolution, which began in the mid-1700s with the popularization of the steam engine, society has been increasing its demand for energy, thus raising pollution levels. Studies carried out by NOAA (the National Oceanic and Atmospheric Administration, the scientific agency connected to the US Department of Trade) have indicated a rise in the concentration of carbon dioxide in the atmosphere on the order of 40% over the past 250 years, mostly owing to the combustion of fossil fuels. That is one of the primary causes of the greenhouse effect, which has been increasing the planet's temperature with potentially devastating effects on biodiversity and, consequently, on the very future of humanity.

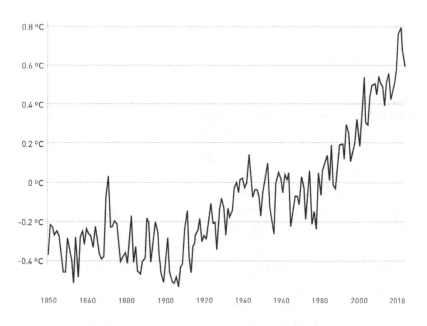

WORLD AVERAGE TEMPERATURE VARIANCE

DIFFERENCE OF LAND AND OCEAN TEMPERATURES
AS COMPARED TO THE AVERAGE FOR 1961–1990 (IN °C)

Figure 16.5. Sources: ourworldindata.org, Met Office Hadley Centre

In just 250 years, our industrial society has caused an average increase in Earth's temperature of between 0.8°C and 1.2°C (1.4°F and 2.2°F). And the effects of an increase of between 1.5°C and 2.0°C (2.7°F and 3.6°F) could be catastrophic and potentially irreversible, according to a report published in October 2018 by the Intergovernmental Panel on Climate Change (IPCC), a body of the United Nations created in 1988. The IPCC is responsible for looking at the scientific aspects of climate change, including its economic, political, and natural impacts; it shared the Nobel Peace Prize with former US Vice President Al Gore in 2007.

The *Special Report on Global Warming of 1.5°C* was signed by 91 authors from 40 countries, and it contains more than 6,000 scientific references. It stated that a global increase of 1.5°C (which, if nothing is done, will be reached between 2030 and 2050) will lead to health problems, a rise in the rate of illnesses such as malaria and dengue, food insecurity, water scarcity, and a reduction in economic growth, among other problems. The sea level could also rise by 80 cm (31 in) by the end of the century, and potentially by nearly 1 m (39 in) in the event of a 2°C (3.6°F) rise, which would cause irreversible instability in Antarctica and Greenland, and a loss of up to 99% of coral reefs. The natural habitats of insects, plants, and vertebrates will shrink by up to half. But avoiding these catastrophes will not be simple: By 2030 carbon emissions would need to fall to around half of 2017 levels, and by 2050 the world would need to become carbon neutral.

Taking into account the pace at which clean energy generation is being implemented around the world—at around 55,000 MW per year, according to estimates by the Carnegie Institution of Science (one of the 23 institutions created thanks to the philanthropy of Andrew Carnegie [1835–1919], an industrialist from the steel sector)—these targets are effectively unachievable, since it would take more than 300 years to implement systems for the nearly 20 TW (terawatts) required. Daniel Schrag, from Harvard University and a climate advisor to former US President Barack Obama, believes that it is possible that we will see an increase of 4°C (7.2°F) or more in average global temperatures within this century. Even the successful use of broad-reaching measures through geoengineering techniques may not be enough—but there is no alternative for humanity but to use all political, economic, and social efforts to control this unprecedented crisis.

Inspired by an exercise developed by the website Land Art Generator, let's consider for a moment the estimate for global energy consumption for 2030, which is close to 800 EJ. This is equivalent to around 219.8 TWh. The average intensity of the solar energy that reaches the atmosphere is approximately 1,360 watts per square meter, according to satellite measurements—and around half of this value reaches the oceans and continents (the rest is reflected back into space or absorbed by clouds). Let's say that, in a year, half of the days are sunny, with around eight hours of natural light, which would result in 1,460 hours of sunlight per year. A solar panel with an efficiency of 20% (the current average) would be able to generate almost 200 kWh per year, per square meter.

Dividing the estimated demand of 219.8 TWh by this value, we can calculate the necessary surface area to meet the global energy demand with solar panels: around 1.1 million square km (425,000 square mi), or 0.7% of Earth's land surface (equivalent to the land area of Bolivia). If we include the surface area of the oceans (since the installation of solar panels over water is possible), we are talking about just 0.2% of Earth's total surface, assuming the efficiency of solar panels does not improve (which is highly unlikely).

SPLITTING THE BILL

Which sectors are responsible for the emission of gases that contribute to the greenhouse effect, such as carbon dioxide, methane, and nitrous oxide? Globally, the energy industry alone accounts for more than half of carbon dioxide emissions and around a third of methane emissions; the agricultural sector emits around 50% of the methane and 70% of the nitrous oxide. In the United States, according to the EPA, in 2018 the segments responsible for the emission of greenhouse gases were transportation (28.2%, especially due to petroleum-based fuels), electricity (26.9%, with nearly two-thirds of generation coming from fossil fuels), industry (22%), commercial and residential (12.3%, especially due to the combustion of fossil fuels for heating), and agriculture (9.9%, largely due to cattle raising). Forests offset this by 11.6%, absorbing more carbon dioxide than they produce.

According to the Emissions Database for Global Atmospheric Research (EDGAR, a joint project of the European Commission and the Netherlands Environmental Assessment Agency), in 2018, of the 38 billion metric tons of CO_2 emitted—versus just two billion emitted in 1900—China accounted for 11.2 billion and the United States for 5.3 billion. Any doubt about the effect of industrialization on the quality of the atmosphere can quickly be dismissed when observing the long-term concentration of carbon dioxide in parts per million (ppm): Up until the First Industrial Revolution, this number had been steady at around 270 ppm; in 2018, we had already surpassed 400 ppm.

GREENHOUSE GAS EMISSIONS
BY SECTOR OF THE US ECONOMY, 2018

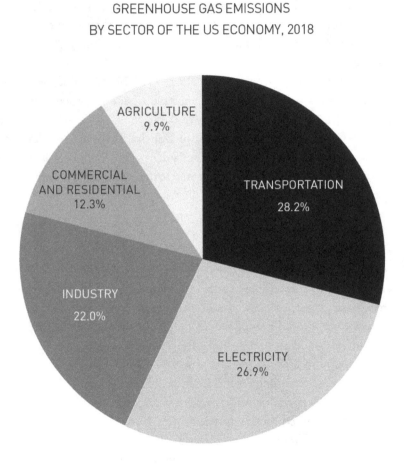

PERCENTAGES MAY NOT ADD UP TO 100%
DUE TO INDEPENDENT ROUNDING

Figure 16.6. Source: Environmental Protection Agency,
Inventory of U.S. Greenhouse Gas Emissions and Sinks

AVERAGE LONG-TERM CONCENTRATION
OF CARBON DIOXIDE IN THE ATMOSPHERE

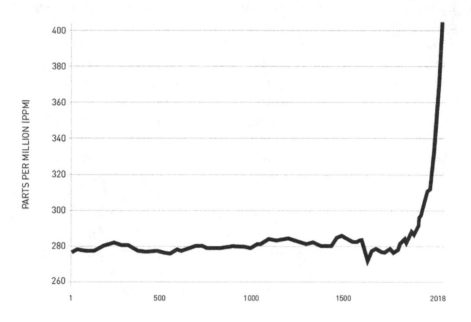

Figure 16.7. Source: University of San Diego—
Institute of Oceanography, Scripps CO_2 Program

The Climate Change Performance Index (CCPI) (published annually by Germanwatch, the NewClimate Institute, and the Climate Action Network) attempts to measure how responsibly the countries that create more than 90% of global greenhouse gas emissions are behaving. Four criteria are used: volume of emissions, renewable energy, use of energy, and climate policies. In the 2019 index, out of a group of 56 countries, (with the European Union counted as a single country), the highest-ranking country was Sweden, followed by Morocco and Lithuania. Countries in last place were Iran, the United States, and Saudi Arabia.

The search for improved energy efficiency and for renewable sources such as solar, wind, wave, and geothermal has become critical both for economic reasons and for reasons of survival. The electronics manufacture and civil construction industries have already prioritized projects and implementations that have a limited environmental impact, using computer programs to simulate a series of variables that affect structures' behavior, their energy consumption, and the waste produced. At the same time, improvements to the infrastructure of urban centers are also being sought, with the installation of LED lights on streets, sensors to reduce consumption, and, as we have discussed, the use of new meters and household energy storage systems that make it possible to use stored energy during peak times (when consumers pay higher rates).

Another change underway touches on the infrastructure of power grids, with the use of distributed energy resources (DER)—the name given to energy sources generally located near the end user (such as wind farms, generators, batteries, and solar panels) and that are connected to the power grid. These types of equipment level out consumption patterns by generating energy that can either be used immediately, stored and then used later at peak hours, returned to the power grid (thus reducing expenses), or even used in the case of a failure in the infrastructure.

Thinking about how to address the issue of energy generation and of the pollutants in our atmosphere, researchers at the University of Antwerp and the University of Leuven, both in Belgium, have developed a device that purifies the air while simultaneously generating energy. Led by Professor Sammy Verbruggen, scientists built a piece of equipment that removes pollutants and converts them into hydrogen using solar light, nanoparticles, and a photoelectric chemical membrane—and it is this hydrogen that can be used as an energy source.

Professor Cary Pint, from the Department of Mechanical Engineering at Vanderbilt University in Tennessee, is coordinating a research project that uses a layer of black phosphorus just a few atoms thick. When the material is bent or compressed at frequencies in line with a person's walking stride, a small electric current is generated. Since its size is on a nanometric scale, it can be incorporated into clothing—in other words, in the future, you may be able to charge your cell phone battery simply by moving.

The use of new materials has become more and more common in several different sectors other than energy: medicine, manufacturing, automotive, and aerospace are just a few examples. Technology applied to materials science is the topic of the next chapter.

17

NEW MATERIALS

IT IS FAIR TO SAY THAT ALMOST EVERY TECHNOLOGICAL EVO-
lution in the history of civilization was tied to the discovery of new elements
and materials. Their importance is so great that archeologists and historians have
even split out the study of ancient societies based on the dominant materials of
the time, using names like the Stone Age, the Bronze Age, and the Iron Age.

The start of the Stone Age has been estimated based on fossils approximately 3.4 million years old that were found in 2010 in Ethiopia and carried
markings made by stone tools. The Bronze Age (circa 3300 BCE –1200 BCE)
exemplified how technology creates competitive advantages for societies who
make use of it: Groups able to smelt copper with tin, arsenic, or other metals
obtained bronze, the hardest and most durable metal (of the time). The third
and last age is that of Iron, from 1200 BCE to approximately 800 CE.

FORGING CHANGE

New materials create entire new industries, and thereby change the course of
societies. Stone, bronze, iron, glass, modern porcelain, and gunpowder led to
significant political, economic, and social developments around the world.

Two of the four great inventions looked on by the Chinese as symbols of their ancestors' sophistication date back to around the ninth century: gunpowder and printing. The other two inventions are the compass (initially used to predict the future starting in the second century BCE and applied to navigation starting in the eleventh century CE) and papermaking (a process documented around the first century). Chinese inventor Bi Sheng (990–1051) created the world's first movable type technology, and four hundred years later German blacksmith Johannes Gutenberg (1400–1468) was the first European to create movable metal printing matrices. This paved the way for massprinted publications by typographers and almost overnight increased citizens' access to information and political, economic, scientific, and religious ideas. Less than a hundred years later one of history's most important revolutions would begin, quite possibly the Big Bang event for the Deep Tech Revolution: the Scientific Revolution, with the publication of *De revolutionibus orbium coelestium* (On the Revolutions of the Celestial Spheres) by Polish astronomer Nicolaus Copernicus (1473–1543). The Scientific Revolution values reason over superstition and experiments as tools to explain the natural phenomena of the world around us. Science and technology began their inseparable journey in an endless feedback loop where science enables the development of better technology, which creates new and better instruments, which enable more accurate and more sophisticated science.

Figures 17.1 and 17.2. Bi Sheng (990–1051), inventor of the world's first movable type printing model. Johannes Gutenberg (1400–1468), who became one of history's most important figures with the invention of the movable type printing system in Europe. Sources: Wikimedia Commons and Shutterstock

Glass lenses started to be perfected in the Netherlands at the end of the sixteenth century and were used to observe the world around us—both things that are very far away (with telescopes) and things that need to be seen very close up (with microscopes). The first patent for the telescope was applied for (but not granted) by Hans Lipperhey (1570–1619) in 1608. The microscope followed a similar path, with different Dutch eyeglass producers claiming to be its inventor (including Lipperhey himself).

In 1800, Italian physicist and chemist Alessandro Volta (1745–1827) created a battery based on copper and zinc. His invention, the precursor of the lithium-ion batteries that power cell phones, tablets, and laptops, was prompted by an amphibian. During the dissection of a frog held down with a brass hook, Italian biologist Luigi Galvani (1737–1798) touched the animal's leg with a steel scalpel, and the leg immediately twitched (the term *galvanism* was coined to describe this phenomenon). While Galvani theorized that "animal electricity" was responsible for the phenomenon, his friend Alessandro Volta suspected that it happened due to the connection between the metals (the hook and the scalpel), a theory known as contact voltage. In reality, the electric energy generated by the voltaic pile—literally a battery made of stacked metals—is the result of chemical reactions, as proven by Englishman Humphry Davy (1778–1829) and his assistant, Michael Faraday (1791–1867), who was also English. Faraday would go on to become one of the most important names in the study of electromagnetism and electrochemistry.

Figures 17.3 and 17.4. Alessandro Volta (1745–1827) and Michael Faraday (1791–1867). Sources: anthroposophie.com (PD-US); Millikan and Gale's *Practical Physics* (1922)

RUBBER AND ENTREPRENEURISM

In the second half of the eighteenth century, the world began to explore the uses of rubber. French botanist François Fresneau (1703–1770), who spent a good portion of his life in French Guiana, was the first to publish a scientific article on the properties of the sap extracted from a specific tree in the region. This sap, called latex, turned into rubber when it hardened. The origin of the word *rubber* comes from the material's ability to erase pencil when rubbing it on paper. This observation is attributed to English chemist Joseph Priestley (1733–1804), who according to several scholars was the scientist who discovered oxygen.

But rubber had a serious problem: It melted in summer and cracked in winter.

In 1834 an American engineer saw some life vests in a store and thought he could improve the valve used to inflate them. But the shop owner suggested he instead work on improving the rubber the vest was made of, and over the next five years he spent all his family's money in search of a solution. Charles Goodyear (1800–1860) solved this problem by inventing the vulcanization process (in honor of the Roman god of fire) after accidentally combining sulfur and rubber, which increased its strength.

Goodyear's story is emblematic for any entrepreneur.

- First, because he made an incorrect diagnosis of the market: He wanted to tackle a problem (the life vests' valves) when the actual problem was the vests themselves.

- Second, he risked everything he had to solve the problem, selling almost all of his possessions (including his children's schoolbooks).

- Third, a major part of his success was owing to a stroke of luck (the accidental combination of sulfur and rubber).

- Fourth, he wasn't able to patent his solution until 1844 (10 years after he started his work on the subject).

- And, lastly, when he died he was once again in debt, this time for the legal fees he incurred fighting violations of his patent.

Figures 17.5 and 17.6. Charles Goodyear (1800–1860), inventor of the vulcanization process. The Goodyear Tire and Rubber Company was founded in 1898 and named in his honor. Henry Bessemer (1813–1898), who applied for a patent for a method to purify an iron/carbon mixture on an industrial scale. Sources: Shutterstock; Mondadori Publishers

Few materials are as closely linked to economic progress as steel; we can find steel in our vehicles and transportation infrastructure, weapons, buildings, and industry. In 1855, English inventor Henry Bessemer (1813–1898) applied for a patent for a method to purify a mixture of iron and carbon on an industrial scale, thus enabling the production of large quantities of steel. Although other ways of doing this already existed—such as in China, for example, since the sixth century—it was Bessemer who made the universal use of steel possible with his technology. Stronger and more versatile than iron by itself, steel and steel products are a constant in essentially all aspects of modern life. It was only after the popularization of the so-called Bessemer process that steel became an economically feasible material—and a similar process of moving toward economic feasibility, this time involving processors, storage, sensors, and smartphones—is once again transforming the economy.

This new era started to be registered, in color, in an article also published in 1855 in *Transactions of the Royal Society of Edinburgh* by none other than James Clerk Maxwell (1831–1879). This Scottish scientist is one of history's most important physicists, having established fundamental principles and equations in the fields of electromagnetism, optics, and thermodynamics. His interest in the process of vision, particularly the way in which we perceive colors, led him to publish the article *Experiments on Colour* and to suggest the three-color method.

The origins of this method, in turn, date back to the early nineteenth century, when the versatile British physicist Thomas Young (1773–1829)—whose interests included music and Egyptology—postulated the existence of structures in the human eye that had sensitivities to specific colors. And in 1850, German physicist Hermann von Helmholtz (1821–1894) determined which portions of the visible light spectrum caused specific structures in the eye to perceive the colors red, green, and blue. In his 1855 article, Maxwell suggested that if RGB (red, green, blue) filters were used to take black-and-white pictures, then overlaying these images would result in the perception of colors.

Figure 17.7. Scottish scientist James Clerk Maxwell (1831–1879), who derived fundamental principles and equations in electromagnetism, optics, and thermodynamics. His interest in the process of vision, in particular in how we perceive colors, contributed fundamentally to the invention of color photography. Source: *The Scientific Papers of James Clerk Maxwell*, edited by W. D. Niven, Dover (New York, 1890) (PD-US)

Black-and-white photography had been presented "as a gift to the world" by the French government 16 years earlier, in 1839. Developed by Frenchmen Nicéphore Niépce (1765–1833) and Louis Daguerre (1787–1851) after studies involving the chemical properties of silver, the only country that did not benefit from this gift was England. That was because Henry Talbot (1800–1877) had independently developed a process similar to Niépce's and, when he heard the news of the new technology from France, he decided to disclose his methods to obtain the rights for it. According to some historians, in retaliation Daguerre's representatives applied for a patent in England for the Frenchman's invention, meaning that licensing costs would need to be paid in England, but nowhere else in the world.

Another discovery from the same year, 1839, also involved silver: the photovoltaic effect. This occurs when the light hitting a material generates a voltage, (which is called the Becquerel effect in honor of the man who discovered it, French physicist and photography scholar Edmond Becquerel [1820–1891]). The real application of this technology was first made possible by English electrical engineer Willoughby Smith (1828–1891), who worked on the development of submarine cables with Charles Wheatstone (who also created the first stereoscopes, as we have seen before).

During his research, Smith verified that the more light that fell on selenium (an element discovered in 1817 by the great Swedish chemist Jacob Berzelius [1779–1848]), the greater its capacity to conduct electricity; he documented this in the journal *Nature* in February 1873, in the article "Effect of Light on Selenium During the Passage of An Electric Current." Three years later, in 1876, Professor William Grylls Adams (1836–1915), from King's College in London, and his student Richard Evans Day determined that selenium could generate an electric current when subjected to light, effectively leading to the creation of photovoltaic cells. The first documented solar panels were created in 1883 by American inventor Charles Fritts (1850–1903) and installed on a roof in New York the following year. Their efficiency was around 1%, which means that only 1% of the light that fell on them was converted to electricity. But the evolution of materials used in solar panels raised this percentage to around 5% in the 1960s, and to around 20% 50 years later.

ON THE SHOULDERS OF GIANTS

One of the key characteristics of technology is its capacity to leverage possibilities—each discovery uses the features and properties of a long line of past innovations. The nature of scientific and technological advancement is recombinant, using different components with specific functionalities to make something new.

One of the biggest rivalries in the history of science occurred in the seventeenth century between two Englishmen: Sir Isaac Newton (1643–1727) and Robert Hooke (1635–1703). The correspondence exchanged between the two is a testament to the fact that even some of the most brilliant minds to ever live were not immune to vanity or pettiness. Their level of antagonism was so great it is said that when Newton succeeded Hooke as president of the Royal Society (one of the most traditional and respected societies for the advancement of science, founded in England in the late 1660s), he ordered the only picture of his predecessor removed from the walls. Over time, Hooke's reputation continued to decline—maybe rightly so, given his temperament, but possibly to a greater degree than actually deserved.

Figures 17.8 and 17.9. Isaac Newton (1643–1727). Robert Hooke (1635–1703). Sources: Godfrey Kneller (1689); Rita Greer (2004)

Rivalry aside, in one of Newton's letters to Hooke, dated February 6, 1675, he wrote words that ended up going down in history: "If I have seen further it is by standing on the shoulders of giants." The origin of this expression seems to lie in the work of French philosopher Bernard de Chartres (1070–1130), and it reflects a concept discussed previously: Scientific and technological advances are cumulative, exponential, and inevitable.

The development of new materials and applications continued to intensify in the twentieth century, with the discovery of superconductivity in 1911 by Dutch physicist and Nobel Prize in Physics winner Heike Kamerlingh Onnes (1853–1926). This technology paved the way for the development of the MRI and a few types of bullet trains. Superconductors are materials that, when cooled below a certain temperature, have no electrical resistance. This means that an electric current applied to them could persist for a virtually unlimited amount of time.

In the 1930s, materials that are now part of the day-to-day lives of billions of people were created in the labs of the giant American firm DuPont. Think neoprene (1930), nylon (1935), and Teflon (1938). Curiously, DuPont had originally been formed in 1802 in the United States to manufacture gunpowder around 1,000 years after its invention in China. In 2017, DuPont and Dow Chemical (founded in 1897, also American) merged, forming DowDuPont, whose 2018 revenue exceeded $85 billion.

TRANSISTORS AND THE MODERN COMPUTERS

Few items better define the twentieth century and the Third Industrial Revolution than the transistor, developed in December 1947 by American physicists William Shockley (1910–1989), John Bardeen (1908–1991), and Walter Brattain (1902–1987) at the famous Bell Labs. The lab, formed in 1925 from the integration of the engineering departments of American Telephone & Telegraph (AT&T) and Western Electric (primary supplier to AT&T), has become one of the world's most important technology development centers. By 2018, nine Nobel Prizes (including one for the invention of the transistor in 1956) and three Turing Awards (created in 1966 by the Association for Computing Machinery as a tribute to Alan Turing) had been bestowed upon scientists whose work originated and was developed at Bell Labs.

The creation of the transistor was spurred by the search for a smaller and more efficient solution (two characteristics that often inspire technological advancements) to replace vacuum tubes, also known as valves, which were the precursors of our digital world. Developed during the first half of the twentieth century, vacuum tubes are generally made up of three components: a cathode, an anode, and a control grid.

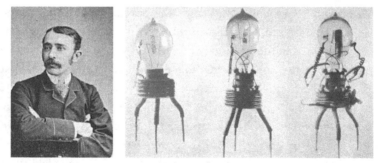

Figures 17.10 and 17.11. John Ambrose Fleming (1849–1945),
English electrical engineer and physicist, inventor of the first vacuum tubes
in 1904. Sources: *Electrical World*, Vol XVL No 10, McGraw-Hill (New York, 1890);
"The Thermionic Valve and its Developments in Radiotelegraphy and Telephony,"
The Wireless Press, (London, 1919)

The word *cathode* comes from the combination of the Greek words *kata* ("down, or negative") and *hodos* ("way"), while *anode* comes from the combination of the words *ana* ("up, or positive") and *hodos*. Put simply, an electric current passes through the cathode, heating it up and releasing electrons, which are attracted by the positively charged pole. As the device's name itself says, this process occurs in a vacuum tube, imposing little resistance on the electrons. The control grid functions as a type of switch: If it is positively charged, the electrons can flow from the cathode to the anode, thus closing the circuit. If it is negatively charged, the electrons are repelled, and the circuit will remain open. This is the principle behind digital electronics, where all information is codified into one of two states: off or on, open or closed, zero or one.

This groundwork led to the development of electronic computers—in contrast to human computers, which is the name given to the people who carried out (or computed) the calculations necessary to solve a given problem. This term was applied to assistants to astronomers back in the seventeenth century, to teams responsible for the compilation of trigonometry and logarithm tables in the eighteenth and nineteenth centuries, to teams that worked during the Second World War on problems linked to nuclear reactions, and to NASA staff involved in the calculations of early manned missions.

The American physicist and professor John Atanasoff (1903–1995) is known as the inventor of the first electronic computer, together with his graduate student Clifford Berry (1918–1963). This computer was called the ABC (Atanasoff-Berry computer) and was developed between 1939 and 1942. Containing around 300 vacuum tubes, it introduced the use of binary digits to represent the information processed. After the ABC, the two most important computers were developed to meet military demands. The first was the Colossus, in England (1943), devised by engineer Tommy Flowers (1905–1998) to decipher encrypted messages from the German High Command of Nazi Germany (which, unlike the troops, did not use the Enigma machine that we discussed in Chapter 4). The second was the ENIAC (Electronic Numerical Integrator and Computer) in the United States (1945), whose initial objective was to calculate the parameters to adjust the artillery's angles of fire, but which, under the influence of John von Neumann (1903–1957), ended up being used in the study of thermonuclear weapons. Colossus used around 2,000 vacuum tubes, but ENIAC needed approximately 20,000, occupying more than 150 m² (1,600 sq ft) of floor space and weighing nearly 30 metric tons.

Figures 17.12 and 17.13. John Atanasoff (1903–1995), inventor of the first electronic computer, known as the ABC (Atanasoff-Berry computer). Sources: Wikimedia Commons and Iowa State University, Mark Richards

Figures 17.14 and 17.15. John von Neumann (1903–1957), who made important contributions in mathematics, physics, and computer science, and ENIAC (Electronic Numerical Integrator and Computer), another example of an important technological advance initially developed for military purposes. Sources: United States Department of Energy (PD-USGov); Paul W Shaffer, University of Pennsylvania

THE CHEMISTRY OF ELECTRONICS

With the advent of transistors, the age of miniaturization of electronics was underway. Shockley and his team used germanium (discovered in 1886 by German chemist Clemens Winkler [1838–1904]) and gold for their first prototype. Both germanium and its replacement, silicon (discovered by the Swede Jacob Berzelius, in 1824), are semiconductors whose capacity to conduct electricity can be controlled. This characteristic enabled the development of the bipolar junction transistor (patented in 1948 and unveiled to the world in 1951).

TRANSISTORS

The operation of the transistor, which is probably one of the two inventions that most profoundly affected the direction of civilization (the other being the steam engine), is conceptually similar to the operation of the vacuum tube in the sense that it works as a gate that allows an electrical current to flow through it or not. Silicon is a chemical element with four electrons in its valence shell (the name given to the outermost layer of the atom, where electrons orbit). In nature, this layer is stable when it contains eight electrons, so whenever possible, atoms seek to bind with other atoms to make up a total of eight electrons in their valence shell. This is what occurs when a silicon atom binds with four others.

Phosphorus was discovered by German alchemist Hennig Brand (1630–c.1710) in 1669, after a laborious process that involved filtering and processing several buckets of urine (an excellent source of raw material for obtaining the element). Its valence shell has five electrons, so when it bonds to silicon one electron is free, thus giving a negative charge to the material (which is then named N). The next element mixed with silicon was boron (discovered in 1808 by French chemists Joseph-Louis Gay-Lussac [1778–1850] and Louis-Jacques Thénard [1777–1857] and, independently, by English chemist Humphry Davy [1778–1829]). With just three electrons in its valence shell, a bond with silicon generates a net positive charge, making the material positively charged (and then it is named P).

The NPN transistor works thanks to the interaction between the excess electrons in the N shell and the missing electrons in the P shell. When the gaps in the P shell are filled, a layer is formed that prevents more electrons from being absorbed—but, as in the case of the vacuum tube, if a positive current is applied to the transistor's base, electrons are able to flow.

LCDs AND OPTICAL FIBERS

Another new material omnipresent in the modern world was developed in RCA's labs by American engineer George Heilmeier (1936–2014) and introduced to the world in 1968: the LCD (liquid crystal display) monitor. Liquid crystals have two characteristics that made their application possible. The first is that they behave as a liquid, filling the receptacle they are in, regardless of its shape. The second is that, unlike liquids, their molecules tend to align (which is a characteristic of crystals). Furthermore, a nearly perfect alignment can be achieved with the use of a small voltage, making the information to be shown either visible or invisible. Monitors, TVs, smartphones, tablets, digital watches, and laptops are a few of the devices that have incorporated this technology. High-resolution liquid crystal monitors, typically lit with LEDs (light-emitting diodes) and relatively new in our day-to-day lives, may be refined in the future with quantum dot technology; these dots emit light at specific frequencies depending on the electrical charge applied.

Then, in 1970, came optical fibers, invented by Corning Inc., an American company specializing in ceramic and glass applications. With the thickness of a strand of hair, these fibers can transmit data in the form of light signals at speeds that had been unimaginable just a few years earlier. Fiber optics revolutionized the telecommunications market, enabling the transmission of vast quantities of data at high speed.

Chapter 14 discussed nanotechnology, whose development was partly owed to the 1985 discovery of fullerene—a versatile form of carbon that looks like a soccer ball and, when laid out in cylindrical format, forms carbon nanotubes. Harold Kroto (1939–2016), from the University of Sussex, along with Robert Curl and Richard Smalley (1943–2005), of Rice University in Houston, Texas, won the 1996 Nobel Prize in Chemistry for its discovery.

Carbon nanotubes—whose popularization is largely due to the efforts of Japanese physicist Sumio Iijima—are lined with one-atom-thick walls, called graphene. Graphene, which is currently the most famous of the so-called two-dimensional materials, has applications everywhere from energy storage and the development of faster electronic components to new water desalination and filtration techniques. Nanomaterials represent one of the most interesting development areas, paving the way for sensors that we can wear 24/7 and smart clothing—clothing that goes beyond its traditional use through the use of technology—a market expected to reach more than $5 billion in 2024 according to MarketsandMarkets Research. Examples include coats that repel water, athletic gear that measures the strain on your muscles, and socks that can measure the pressure each part of your foot receives while walking or running.

METAMATERIALS

The combination of traditional elements in specific sizes and shapes has set the stage for the area of metamaterials, which can affect electromagnetic waves and eventually make possible products that were once found only in science fiction, such as cloaking devices (that allow an object to become partially or entirely invisible to the naked eye). Aerogel—a product where the liquid portion of a traditional gel is replaced by a gas—has been under study as an excellent insulator due to its low thermal conductivity, with important implications for industries such as construction and aerospace. The origin of aerogel, according to several sources, was a bet between University of Illinois chemistry professor and chemical engineer Samuel Kistler (1900–1975) and his colleague Charles Learned. The goal of the bet was to take all of the liquid out of a gel and replace it with a gas, without the material's structure coming apart. In 1931, Kistler won the bet.

Through the use of specialized computer programs (that can precisely simulate the behavior of atoms, molecules, and compounds subject to different physical and chemical processes), the development of new materials has become even more flexible and less expensive. Just as many areas of the pharmaceutical industry use simulators to analyze the behavior of new drugs in the human body, thanks to advances in the area of big data (the field that studies the processing of vast quantities of data, and the topic of the next chapter), new materials are created inside computer processors before becoming part of the physical world.

BIG DATA

EVERY FOUR YEARS, THE GENERAL CONFERENCE ON WEIGHTS
and Measures takes place in Sèvres, around 10 km (6.2 mi) from the center of
Paris. Attendees define all aspects of the metric system, which is used in the
daily lives of the majority of the world's population (the only exceptions are
Liberia, Myanmar, and the United States). The first of these events was held
in 1889, and some of the most commonly used units in the world of personal
computing, such as mega, giga, and tera, were decided upon at the 1960 conference. In 1975 and 1991, additional prefixes were created to support the
discussion of even larger orders of magnitude: peta and exa in 1975, and zetta
and yotta in 1991.

When we convert from one prefix to the next, we are increasing values by a
factor of a thousand—for example, one peta is 1,000 times larger than one tera.
It should be noted, however, that there is a difference between the factor used
in the decimal system, which is 1,000, and the one used by the binary system,
which is $2^{10}=1,024$ (in this case the prefixes would be, for example, kibi rather
than kilo, mebi rather than mega, gibi rather than giga, and so on).

FROM BITS TO YOTTABYTES

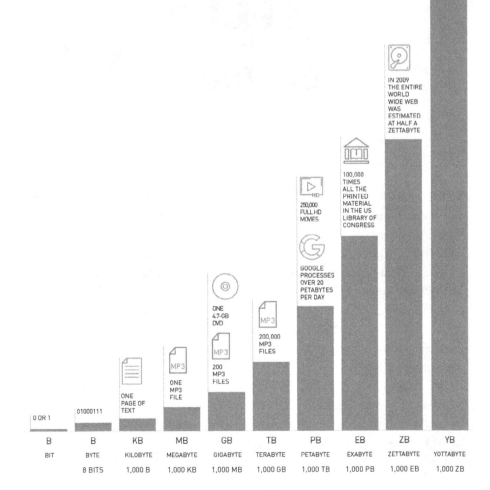

Figure 18.1. Sources: Wikipedia, others

MACHINE TALK

To offer some perspective on the quantity of information that is constantly being created, here is what was happening on the Internet each minute according to Domo, a software company specializing in business intelligence software. In September 2020, the tally was around 6,600 packages shipped by Amazon, 500 hours of YouTube videos uploaded, 147,000 photos uploaded on Facebook, 404,000 hours of content watched on Netflix, 41 million WhatsApp messages sent, and $1 million spent in online purchases. Again, all of this happened in just a period of 60 seconds.

But we also need to consider the universe of data generated by the growing number of devices connected through the IoT. With the colossal volume of information that is constantly being created and updated, we need to develop tools and technologies to analyze and interpret the data. This relentless generation of data is one of the most significant characteristics of the widely connected and integrated world to come.

According to estimates from Cisco, between 2017 and 2022 Internet data traffic is going to grow 26% per year—from 122 exabytes per month to nearly 400. It is estimated that, counting only mobile devices, data traffic will see a sevenfold increase by 2022, representing an annualized growth of 46%. Devices that communicate directly among themselves, with no human intervention (a modality known as M2M, or machine to machine), accounted for just over six billion connections in 2017, and by 2022 this number is set to exceed 14.5 billion. And it is not only the number of connections that is going to increase: In 2017, M2M traffic was around four exabytes per month. This volume is projected to hit 25 exabytes per month by 2022.

Cisco's report also states that by 2022, virtually 50% of M2M connections will be inside our homes (as part of home security and automation equipment, for instance). The next-largest source of these types of connections will be in workplaces, followed by three segments we have already discussed: health, cities, and vehicles. This also gives us an idea of the type of data machines will be exchanging over the coming years, such as medical data (including videos and images and health indicators) and data from navigation systems.

Over the coming years we will witness a significant increase in the quantity of data generated and transmitted—by us, our friends, relatives, co-workers, and by the machines around us. Each email, message, photo, video, audio, or piece of data contributes to one of the fundamental challenges we will need to face thanks to the technology that is omnipresent in our lives: How to extract

IP TRAFFIC (EXABYTES PER MONTH)

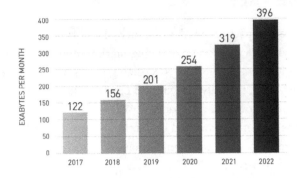

M2M TRAFFIC (EXABYTES PER MONTH)

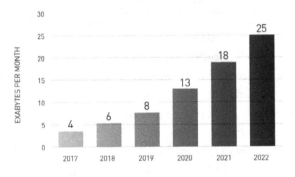

M2M CONNECTIONS, BY INDUSTRY (IN MILLIONS)

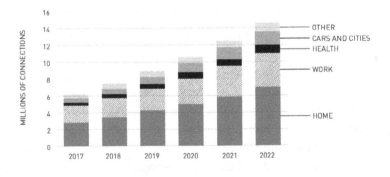

Figure 18.2. IP and M2M traffic. Source: "Cisco Visual Networking Index: Forecast and Trends, 2017–2022"

the relevant data out of that massive volume of bytes? How will we use the unimaginable quantity of valuable information that is being produced every day, and do so in an intelligent manner? This is the challenge that *big data* seeks to address.

THE FUTURE IN DATA

The term *big data* became increasingly well-known toward the end of the 1990s, and its characteristics are usually associated with the five Vs: volume, variety, velocity, variability, and veracity. In 2016, three Italian researchers (Andrea De Mauro, Marco Greco, and Michele Grimaldi) published an article in which they defined big data as "the information asset characterized by such a high volume, velocity and variety to require specific technology and analytical methods for its transformation into value."

Big data techniques work by using all available data (not just a subset thereof), and they are able to work with information that, in computer science jargon, is known as *non-structured*. This means that the data used does not need to be of the same type, nor organized in the same way, and it is often processed in a distributed manner, in parallel (i.e., simultaneously, by several processors). The opportunities for extracting value from information generated by different business segments are potentially worth on the order of hundreds of billions of dollars, and virtually all sectors—both private and public—that produce data can benefit from the analyses performed by business intelligence systems.

If we take the example of the comments posted on social media about a sporting event, we will have different types of data (text, images, videos, audio), generated by multiple sources (spectators, fans, reporters, players) and produced over the course of seconds or days. With the proper analysis of this data, it is possible to perform value extraction for consumers and brands.

During the next three minutes—the time it will take you to read the next few paragraphs—more than 10 million Google searches will have been performed, around 1.5 million new tweets will have been posted on Twitter, and 1 million new stories will have been posted to Instagram. According to IBM, every day more than 2.5 quintillion bytes are produced—the equivalent of 2.5 million terabytes. As we have seen, with just 10 terabytes it would be possible to store the content of all the books in the US Library of Congress, the world's largest library.

YOUR DATA OR *OUR* DATA?

The recommendations you see when making online purchases on Amazon would not be possible without the use of techniques for analyzing extraordinary quantities of data. Computer systems cross-reference information from your purchasing history with that of other consumers and consider factors such as demographics, season, location, and previous searches. Something similar happens when you want to select a movie on Netflix when it suggests titles you are likely to be interested in, based on your selections and ratings and those of other users.

On Christmas Eve 2013, Amazon took things a step further. With patent number US 8,615,473, the company registered *Method and system for anticipatory package shipping*, indicating its plans to ship consumer goods to its clients before they even place their orders. The use of data combined with smart algorithms enables the company to predict, with some degree of success, what a consumer's next order will be.

Google also uses the collection of searches performed globally to attempt to guess what we're all searching for: Using the first few letters of your search, the algorithms look through the most frequent searches performed, and they auto-suggest the question you may be asking. (Google tries to avoid suggestions with negative or defamatory content, although this is not always possible.) The data provided by the users themselves is a powerful asset for Google, and that fact, to a large extent, makes it one of the world's most valuable companies, with a market value of more than $900 billion at the end of 2019.

In November 2008, scientists at Google partnered with the US CDC (Centers for Disease Control and Prevention) to publish an article in *Nature*, entitled "Detecting influenza epidemics using search engine query data." The work compared historical information on epidemics against approximately 50 million words (both related and unrelated to influenza) that appeared most often in the more than three billion searches carried out by users every day. Then, after testing nearly half a billion mathematical models that correlated the searches to epidemics, the system identified the 45 terms that best fit the data. Thus, based on the information provided by the users through their searches, Google

continued

---

gained the ability to identify, with a high degree of precision, where a flu epidemic was occurring.

In April 2020, during the COVID-19 pandemic, Google and Apple partnered to develop a contact tracing app to anonymously keep track of nearby phones. If one of the owners on this list of phones was diagnosed with the virus, alerts would be sent to people who had been nearby.

All these examples speak to the importance of data in our society—and like every valuable asset, it has to be protected and used wisely. Since the beginning of the 2010s, countries have been writing legislation on the matter, and arguably the most famous piece is the General Data Protection Regulation (GDPR). Written in 2016 by the European Parliament and the Council of the European Union, its fundamental objective is to make sure individuals in the European Union and in the European Economic Area have control over their personal data.

The balance between making sure our data is serving us—which may only be achieved by sharing it—while at the same time preserving our rights to privacy is a complex issue, but it is an inevitable discussion in every society around the world.

In 1979, American historian Elizabeth Lewisohn Eisenstein (1923–2016) published a book titled *The Printing Press as an Agent of Change*. She proposed that its invention by German-born Johannes Gutenberg (1400–1468) in the fifteenth century created the necessary conditions for the Renaissance and the Scientific Revolution—which, together with the First Industrial Revolution, could be considered as the origins of modern society. According to Eisenstein, between 1453 and 1503 approximately eight million books were printed—more than all the written material produced in the nearly 5,000 years of civilization up until then. And the speed at which more data is being created is mind-blowing: In the early 2020s, it is estimated that each one of us produces almost two megabytes of data—per second. And according to a World Economic Forum post using data from multiple sources, by 2025 we will see the creation of 463 exabytes of data every single day.

One of the main challenges of big data techniques is how to turn massive amounts of data (stored in all different formats) into useful information. This is generally done by analyzing the correlations among thousands of variables, without knowing beforehand which ones will turn out to be relevant. The algorithms extract the necessary recommendations from these correlations. However, it is important to note that correlation does not imply causation. Simply because two variables move in a similar fashion does not necessarily mean that one explains the other. In 2015, Tyler Vigen published *Spurious Correlations,* where he presents several examples of variables that are highly correlated, but that clearly have no cause-and-effect relationship. Consider the divorce rate in the state of Maine and the per-capita consumption of margarine between 2000 and 2009: There is a 99.26% correlation between these two things, but there is clearly no relationship between one and the other.

The selection of a large number of variables is typical of big data systems. Imagine, for example, an application in the area of public security. It is possible to generate recommendations that help prevent crime by analyzing data collected by police on types of crime and overlaying them with the city's calendar of events, weather, and movement of people. Moreover, computer systems are already being used that autonomously assess, based on historical data, whether a given arrestee should be conditionally released, the bail amount, and even the duration of a sentence. How effective, or even fair, this will turn out still remains to be seen—but as we have already seen in Chapter 4, particular attention must be paid to the bias contained in many (if not most) training data sets.

The smart use of data could also provide benefits in terms of maintaining the infrastructure of a large city. In their 2013 book *Big Data: The Essential Guide to Work, Life and Learning in the Age of Insight,* Viktor Mayer-Schonberger and Kenneth Cukier describe how New York's gas and power company, Consolidated Edison (Con Ed), used technology to reduce the risk of manhole explosions.

The company's objective was to predict which manholes were going to show problems so that corrective measures could be taken. Manhattan has more than 50,000 manholes and more than 150,000 km (93,000 mi) of cabling, so determining exactly which ones should be prioritized for inspection is a complex task. Researchers from Columbia University, led by Cynthia Rudin (professor of computer science, electrical engineering, and statistics at Duke University in North Carolina) tabulated data that had been collected by maintenance teams beginning at the end of the nineteenth century and correlated the data to incidents. Using more than 100 variables to make their predictions, testing indicated that the model was able to correctly predict over 40% of the manholes that would present problems.

Another public utility using the power of information that is constantly made available by users is the United States Department of Education, which in 2012 published a report entitled *Enhancing Teaching and Learning Through Educational Data Mining and Learning Analytics.* With the increasing popularity of online courses, it is possible to monitor students' behavior and performance, assist in their development, and provide input for course providers to adjust content.

OUR DIGITAL TRACKS

The word *routine*, which is French in origin, comes from the word *route* (meaning road). Literally, routine means the road normally taken, and, figuratively, it is the habit of doing something in the same way. So, both literally and figuratively, a routine is a repetitive behavior in our lives. This repetition—our habits—creates a structure within which we feel security and familiarity and can become automatic in our minds within just a few weeks. In 2006, researchers at Duke University reported that approximately 45% of our daily behavior is a repetition of some type.

In an age where we leave digital tracks wherever we go—which websites we visit; which e-commerce products we are interested in; which sports, movies, and shows we watch; what music style we listen to; which roads we travel on—more than ever, our habits can be watched, quantified, and measured. Smart cities, for example, can make use of traffic information to plan new roads, reverse the direction of certain streets at specific times, and reduce bottlenecks, thus decreasing pollution. Your Internet browsing sessions can be interrupted by ads geared specifically to you—for example, an ad for the world parachuting championships, based on a past purchase or a trip connected to parachuting locations you may have made.

The processing power that is now available enables corporations to detect subtle changes in consumers' habits—and this has become a significant business opportunity: A graduation, change of city, marriage, pregnancy, or divorce are all indicators of possible changes in consumption habits. A story that reflects this new dynamic is told by American journalist and writer Charles Duhigg in his 2012 book *The Power of Habit.*

In 2018, the US-based retailer Target had nearly 2,000 stores nationwide. Attuned to the habits of its customers and seeking new opportunities to build loyalty, the company's analysts were tasked with detecting when a woman became pregnant, using data available in their systems. Pregnancy significantly modifies consumption habits, so the earlier Target learns of this event, the faster it can act with relevant offers.

The team of technicians analyzed historical purchasing patterns of women who signed up for the baby registry on the company's website, letting big data techniques detect correlations that would show products with a likelihood of indicating a pregnancy. Among the 20 products the system yielded were moisturizing creams and dietary supplements. Based on the dates of these purchases, Target identified not only their pregnant customers, but also their stage of pregnancy. The next step was to introduce a program that offered products recommended specifically for each trimester.

But that wasn't the end of the story told by Duhigg. He recounts the story of a father who went into one of the outlets demanding to speak with the manager, furious that the retail chain was sending his daughter (who was still in high school) coupons for baby clothes and cribs. The angry father asked the company to stop, saying it would encourage his daughter to get pregnant. The issue was escalated inside the company, and a few days later the father was contacted by a representative. During their conversation, the father ended up telling the agent that after his visit to the store, he had learned that his daughter actually was pregnant, and that the offers were therefore appropriate.

WHAT BIG DATA TECHNOLOGY IS LOOKING FOR

The primary objective of big data technology is the extraction of recommendations based on a vast sample set. Sometimes all samples available are taken into consideration; in others, just a subset is used. Imagine a traffic engineer needs to analyze traffic patterns in a certain part of a city between 5:00 and 7:00 p.m. Using traditional techniques, samples of routes taken by a few vehicles would be used to support planning for any prospective actions. But with big data, the routes taken by *all* vehicles can be analyzed. In statistics this is known as n=N. The number of samples (represented by n) is equal to the total number of events (N).

An important characteristic of big data techniques is that understanding the why of a certain phenomenon is not the most important thing. The recommendations obtained are at times counterintuitive, but they work empirically—since the data shows the result unequivocally. This is a consequence of the growing complexity of information systems, something that is set to increase over time. The sensors surrounding us in our daily lives, the constant gathering of data by smart cities, and the development of data capture techniques (such as cameras equipped with facial recognition systems) provide big data algorithms (whether based on artificial intelligence techniques or not) with the raw materials they need to generate their predictions. This field is known as predictive analysis.

Predicting the contours of a specific phenomenon is an advantage that many businesses are willing to pay for. What will be the effect of a certain marketing initiative? How long will it take for one of the fleet's delivery trucks to start showing problems? When will one of the machines on an assembly line stop working? What is a patient's risk of having a major medical complication in the near future? In his 2013 book, *Predictive Analytics: The Power to Predict Who Will Click, Buy, Lie, or Die,* American professor Eric Siegel cites examples of how different companies use this technique.

One example had to do with how Facebook selects the order of news items it shows its users, seeking to maximize usefulness and revenues on its platform. Another example showed that some health insurance companies are now able to estimate certain deaths more than a year in advance, allowing them to begin counseling and support services for families.

The security of this massive quantity of data—one of the most important assets of the Fourth Industrial Revolution—has become a central issue for individuals, families, communities, governments, and businesses to deal with. How can we protect data and still guarantee appropriate levels of access while respecting privacy and neutrality? How can technology act to solve a problem that technology itself created? Cybersecurity will be our next topic.

CYBERSECURITY AND QUANTUM COMPUTING

ADVANCES IN CONNECTIVITY AMONG COMPUTERS, SMART-phones, tablets, vehicles, and other devices used daily by billions of people have been relentless and powerful. These successes have brought information front and center in modern society's value chain. Data—the raw material of the Fourth Industrial Revolution—is now one of the most precious commodities that individuals, research institutions, industries, governments, and organizations possess. As we saw in Chapter 18, we live in a world where the production, transmission, and storage of an ever-increasing volume of data is inevitable.

Companies exchange information every day on processes, patents, inventions, studies, clients, markets, strategies, hirings, firings, and promotions. Our digital DNA—passwords, preferences, purchasing history, favorite shows, financial status, photos, videos, documents, and presentations—is stored in a complex infrastructure spread over the globe in a vast system of equipment and programs. For security and privacy reasons, the protection of this data plays a central role in an interconnected, global society that is completely dependent on an uninterrupted and reliable flow of information.

THE FIRST HACKER

The history of unauthorized entry into systems (known commonly as *hacks*) started more than a hundred years ago, when the electromagnetic transfer of information—one of the pioneering technologies developed to interconnect segments of society—was still in its early days. In yet another irony of the history of science, one of the world's first hackers was an inventor and professional magician: Englishman Nevil Maskelyne (1863–1924).

One summer day in 1903, Guglielmo Marconi (1874–1937) was in Poldhu, in Cornwall, England, getting ready to send a wireless long-distance transmission of Morse code to the auditorium of the Royal Institution of Great Britain, in London, around 450 km (280 mi) away.

Marconi was one of the pioneers of the development of applications for radio waves, which are one of the many types of electromagnetic radiation found in the universe (such as microwaves, infrared, and x-rays). This radiation, which is produced by objects with an electric charge, is one of the four basic forces of nature, together with gravity (which determines that objects with mass or energy attract each other), weak force (present among subatomic particles, such as quarks and leptons), and strong force (which is basically responsible for keeping atoms together).

But before Marconi could send his demonstration message, the team in London received another message: the word *rats*, repeated several times, followed by excerpts from Shakespeare that had been modified to insult and provoke Marconi.

The hacker completed his transmission just before Marconi's transmission was to take place, and Marconi's went off without any further hitches. But since one of the main advantages promised by the new technology was the privacy of messages, which was underscored in an interview between Marconi and the English newspaper the *St. James's Gazette* (founded in 1880 and merged into the *Evening Standard* in 1905), the hack put the premise in question. A few days later the London newspaper *The Times* (founded in 1785) published a letter from the architect of the prank, Englishman Nevil Maskelyne (1863–1924). Interestingly enough, Maskelyne oversaw an important project undertaken by human computers (please refer to Chapter 17) to make the determination of longitude at sea easier (through the calculation of tables of lunar distances).

In his 2001 book *Wireless*, Sungook Hong explains that Maskelyne, a magician with a particular interest in wireless communication (which was useful for his tricks), like so many other inventors, had stumbled upon the generic patents applied for by Marconi, which were meant to prevent others from making developments in his areas of interest. Obviously, Maskelyne was not Marconi's biggest fan. In his letter to *The Times*, he justified the hack as something that had been done "for the public good," with the aim of exposing and correcting the security flaws of this type of communication—a justification still used today by hackers around the world.

Figures 19.1 and 19.2. Guglielmo Marconi (1874–1937), Italian inventor, and British inventor and magician Nevil Maskelyne (1863–1924). Sources: United States Library of Congress's Prints and Photographs division, digital ID cph.3a40043 and newscientist.com (image: RI)

Maskelyne had been hired two years earlier by the Eastern Telegraph Company, which managed the submarine cables that connected the United Kingdom to Indonesia, India, Africa, South America, and Australia, and which could have seen its business simply disappear with the emergence of the wireless transmission that Marconi was working on. In fact, scientific progress can dramatically impact even well-established businesses (remember Blockbuster and Kodak).

But Maskelyne was far from being Marconi's only detractor. To carry out the first transatlantic radio transmission (completed on December 12, 1901, between Poldhu and Signal Hill, in Newfoundland, Canada), the output of the transmitters needed to be boosted. This task was assigned to John Fleming (1849–1945, Figure 17.10), who never received credit for effectively creating the first large radio transmitter (nor the 500 shares in Marconi's company that had been promised to him).

HACKERS AND CRACKERS

The word *hack* (the root of the word *hacker*) was already a few hundred years old before it became associated with attacks on computer systems. Back in the thirteenth century, in German (*hacke*) and Nordic languages such as Danish (*hakke*), the precursors of the English word were associated with cutting something up in a rough way, with many strikes from a sharp object. In the eighteenth century the term started to be used in reference to someone who was employed to carry out routine tasks, and at the beginning of the twentieth century it started to be used as a synonym of attempt.

Meanwhile, the word *hacker* became associated with innocent experiments that curious engineers carried out on different types of equipment. At a meeting of a student organization at MIT in April 1955, the minutes of the Tech Model Railroad Club (which we wrote about in Chapter 8) note that a certain Mr. Eccles requested that everyone who was working, or *hacking* on an electrical system, "turn the power off to avoid fuse blowing." Over time, the word took on an unjustified reputation for being associated with negative or destructive intentions. The correct term was listed in 1975 in the *Jargon File* (a compilation of the terms used by the pioneers of modern information technology at ARPANET, Stanford, Carnegie Mellon, MIT, and other places) —not *hacker* but *cracker.*

Regardless of terminology, the global interconnectivity among devices has exponentially increased targets available for potential attacks. The overall budget allocated to cybersecurity by governments and companies continues to increase to avoid the material and financial losses caused by a successful hack.

The size and importance of the information-security market—including applications, cloud, data, infrastructure, network, and services, among others—will probably continue to grow for the foreseeable future. According to data from multiple sources compiled by AustCyber (an independent, not-for-profit funded by the Australian government), global cybersecurity spending will reach about $190 billion in 2025, up from about $85 billion in 2015 and $130 billion in 2020.

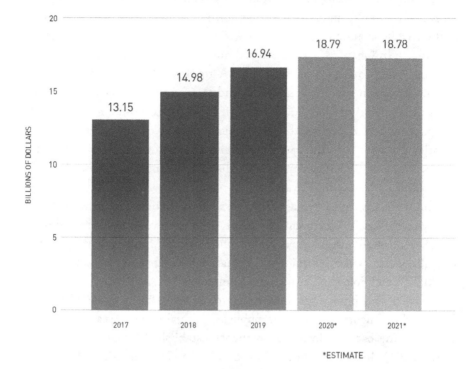

US GOVERNMENT'S BUDGET
FOR CYBERSECURITY

Figure 19.3. Source: US Congressional Budget Office

With the advent of the IoT, any object that is connected to a network becomes a target. Cars; planes; industrial systems; logistics and control equipment; public utilities such as gas, water, and power; hospitals; and even household appliances can all be vulnerable.

Cybersecurity experts are highly valued professionals, and the most prominent consultants can charge thousands of dollars per day. Those who work to detect security flaws and vulnerabilities are known as *white hats*, and their mission is to simulate hacks using multiple techniques so that their employers can improve their defense systems. They have a large arsenal of tools at their disposal, employing methods such as *backdoor*, *DoS* (denial of service), *spoofing* (masking one's own identity), and *phishing* (seeking to obtain information directly from authorized users).

SCHRÖDINGER'S BIT

Cryptography, whose root comes from the Greek word *kryptos* ("hidden") and *graphia* ("written"), underlies the Internet. It enables us to exchange information while maintaining our privacy, thanks to the use of mathematical operations. In Chapter 12 we discussed how pairs of public and private keys play this role: To send a message from A to B, sender A needs to encrypt the message with the public key of B, and to read the message, addressee B must use their private key. Analogously, if B wishes to send a message to A, the public key of A needs to be used, which can only be deciphered with the associated private key. You can imagine the public key playing the role of a padlock, locking the information—and this information can only be disclosed with the use of a specific, private key.

But we can't guarantee that it is impossible to obtain the private key based on the public key. With current processors, based on two states (zero and one), the algorithms are just too slow. But a technology still in its early stages could potentially change the dynamics of information we exchange around the digital world. That technology is the quantum computer.

In 1975, Polish physicist Roman Ingarden (1920–2011) wrote "Quantum Information Theory," which was published the following year. His paper expanded the bases of information theory established by Claude Shannon (1916–2001) in his historic 1948 article "A Mathematical Theory of Communication" into the realm of quantum mechanics, which had emerged in the twentieth century spurred by inconsistencies classical physics was unable to explain. The conclusions reached by scientists such as Albert Einstein (1879–1955), Max Planck (1858–1947), Niels Bohr (1885–1962), Werner Heisenberg (1901–1976), and Erwin Schrödinger (1887–1961) regarding the nature of the phenomena that occur in the world of atomic and subatomic particles not only changed our understanding of how the universe works, but they also scientifically proved concepts that were not intuitive and that challenged logic.

Figure 19.4 and Figure 19.5. Claude Shannon (1916–2001), author of the historic article "A Mathematical Theory of Communication," which was fundamental to the technological advance that powers the modern world. Roman Ingarden (1920–2011), author of the article "Quantum Information Theory." Source: Oberwolfach Photo Collection, Konrad Jacobs and Wikimedia Commons

In 1980, American physicist Paul Benioff's article in the *Journal of Statistical Physics* proved that it was possible (in theory, at least) for quantum computers to exist. The following year, Richard Feynman (1918–1988), in one of his famous talks, indicated that it is not possible to simulate quantum systems with the use of classic computers. This is because (going back to one of the counterintuitive and seemingly illogical concepts proven by quantum mechanics) components of a quantum system can exist in a state of superposition, meaning in more than one place at the same time, with their location described by probabilistic functions. In other words, the act of *observing* the event is what ends up *defining* it.

To attempt to explain this fundamental property of the quantum world, in 1935 Austrian physicist Erwin Schrödinger wrote about what later became known as a thought experiment called Schrödinger's cat. A cat is placed inside a closed box, along with a vial of poison and a radioactive item. If radioactivity is detected inside the box, the vial breaks and the poison is released. According to quantum mechanics, as long as the box remains closed, the cat is both alive and dead. When the box is opened and the cat is observed, the superposition ends and one of the two states (in this case, alive or dead) is revealed (i.e., the observation of the item ends up defining it).

Quantum computers are based on this property, and their applicability in areas such as cryptography, optimization, and simulation of complex systems is the subject of studies (and investment) around the world. Unlike traditional computers, which use bits (binary digits) with a value of zero or one, quantum computers use qubits (quantum bits), which exist in a superposition of states.

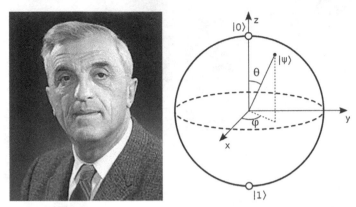

Figures 19.6 and 19.7. A qubit is usually represented using the Bloch sphere, named in honor of Swiss American physicist Felix Bloch (1905–1983). Bloch was the first director-general of CERN (the European Organization for Nuclear Research), where in 1989 English computer scientist Tim Berners-Lee developed the HTTP protocol, effectively creating the World Wide Web. Sources: Stanford News Service and Smite-Meister

PRIVATE AND PUBLIC KEYS

Let's return to the subject of obtaining a private key by using a public key. If this is possible, someone in possession of an intercepted message and the public key used to encrypt it would be able to read it. As difficult as it is to try to reverse-engineer the calculation (i.e., to obtain the private key by using the public key), someone could try to find the private key by carrying out an exhaustive trial-and-error process. The time this would take current computers to accomplish is too long to present a real risk, but the application of Peter Shor's algorithm in quantum computers could change this.

While working at Bell Labs, American mathematician Peter Shor developed an algorithm specifically for quantum computers to determine prime numbers, that when multiplied, would result in a given value. As we have seen, the core of the RSA encryption algorithm, which is used globally in Internet communications, depends on the impossibility (for now) of this factoring operation being executed within a reasonable time frame.

One example of a scheme that wouldn't be broken by Shor's algorithm was created by American mathematician Robert McEliece in 1978; it applies random-ization to the encryption process. This technique is an example of *post-quantum cryptography*, a set of techniques that, if quantum computers become economically feasible, could restore privacy to online communications.

DIGITAL THREATS, REAL LOSSES

In a progressively more connected world, the expansion of the Internet and the falling price of sensors are making it possible for everything and everyone to have an address on the web. While the benefits of this architecture are obvious, with increases in efficiency, access to information, and convenience, the risk of hacks on online systems has never been so great.

We know that in the 1980s, large company systems were already being tar-geted by hackers such as Ian Murphy. To reduce his telephone bill (at the time, the only way to connect to other computers was over phone lines), he hacked into AT&T's system and changed the billing programs, making daytime rates the same as evening rates, which were cheaper. Some say that one of the characters from the 1992 movie *Sneakers*, directed by Phil Alden Robinson, was based on Murphy.

In June of 1990, Kevin Poulsen took control of the telephone lines at the LA radio station KIIS-FM, to ensure he would be the 102nd caller and thus win a Porsche 944. He was arrested by the FBI, sentenced to five years in jail, and banned from using computers for three years after his release. In 2005, he became an editor for *Wired*, an online and print magazine focusing on the impacts of technology on society.

In theory, any item that is connected to the Internet—from an ATM machine to a car, a pressure sensor to a purchase order management system—is susceptible to a hack, because there is always a digital path back to the item. In the 1996 movie *Mission: Impossible* (based on the American TV series that aired between 1966 and 1973), agent Ethan Hunt (played by Tom Cruise) is tasked with breaking into a building to access a computer with information so sensitive that it is not connected to any networks. The creativity and sophistication of hackers is so great that no way exists to provide an absolute guarantee of the security of anything that is connected. Spoiler alert: Hunt and his team manage to steal the information, further demonstrating our data is constantly vulnerable.

A 2007 study at the School of Engineering at the University of Maryland in the United States, indicated that at that time, computers with Internet access were being hacked on average once every 39 seconds. And this only considered as hacks the brute force attacks, where usernames and passwords are randomly tested on thousands of computers until a system can be broken into. Usernames such as *root*, *admin*, and *adm*, and passwords like *123456* or that match the username are still very common and present a high risk for hacking.

TOTAL VALUE AT RISK FROM CYBERCRIME
ESTIMATED AT MORE THAN $5 TRILLION
BETWEEN 2019 AND 2023.

[PONEMON INSTITUTE AND ACCENTURE, 2019]

ONE OUT OF EVERY 131
EMAILS CONTAINED SOME
TYPE OF MALWARE.

[SYMANTEC, 2016]

IN 2007, HACKERS
EXECUTED AN ATTACK
EVERY 39 SECONDS.

[CLARK SCHOOL, UNIVERSITY OF MARYLAND]

IN 2017, MORE THAN 20% OF CYBERATTACKS
CAME FROM CHINA, 11% FROM THE US,
AND 6% FROM RUSSIA.

[SYMANTEC]

IN 2016, AN ESTIMATED
4,000 RANSOMWARE ATTACKS
OCCURRED DAILY.

[FBI]

DOWNTIME COSTS WERE MORE
THAN 20X GREATER THAN THE AVERAGE
RANSOMWARE DEMAND.

[DATTO, 2019]

Figure 19.8. Sources: FBI, others

THE REAL DIGITAL THEFTS

During virtually the entire history of civilization, one of the prerequisites for carrying out a theft has been the presence of the offenders at the scene of the crime. The stolen objects were always part of the physical world—gold bars, paper money, equipment, jewels, vehicles—and there was always an objective value associated with them. The thieves had to have a plan to access the object, a way to remove it from its original location, and the means to transport it to a new location. From there, they would try to modify, resell, copy, or spend the clumps of atoms they had taken unlawful possession of.

But the world has been accelerating the transformation of atoms into bits. Targets of cybercrimes can be thousands of miles away from the perpetrators, who no longer use weapons, getaway cars, or explosives. These thieves are computer systems specialists fluent in techniques ranging from the simplest to the most sophisticated, and their aim is to access information—to copy it, disclose it, modify it, destroy it, or prevent the legitimate owners from accessing it. This information, stored as bits, could represent anything: photos, movies, money, security codes, maps, documents, patents, spreadsheets, presentations, new programs, new games, or genetic codes.

Like all robberies, the modern version starts with a search for vulnerabilities. The hacker uses tools programmed to analyze the system they wish to break into in search of unprotected entry points via brute force methods or more sophisticated techniques that monitor Internet data traffic in search of specific information.

The World Economic Forum's *Global Risks Report* for 2018 listed cybercrime as one of the most significant risks in terms of its likelihood of occurrence. Considering the growing dependence by individuals, governments, and businesses on technology in general, the losses caused by this type of event can be devastating—and, sadly, there is no shortage of examples.

During the negotiations in 2016 for the sale of online services provider Yahoo! to telecom giant Verizon (both US companies), Yahoo! disclosed that it had undergone the largest data leak recorded up to that time: The emails, names, dates of birth, and passwords of no fewer than three billion users had been compromised in at least two hacks between 2013 and 2014. After this information was revealed, the final sale price was reduced by approximately $350 million.

Companies from a wide range of sectors have also been the victims of large-scale hacks. This includes e-commerce site eBay (145 million users affected), retail chain Target (110 million customers), ride-hailing company Uber (57 million users and 600,000 drivers affected), JP Morgan Chase (76 million households and seven million small businesses), and the Sony PlayStation Network (77 million users affected and approximately $170 million in losses).

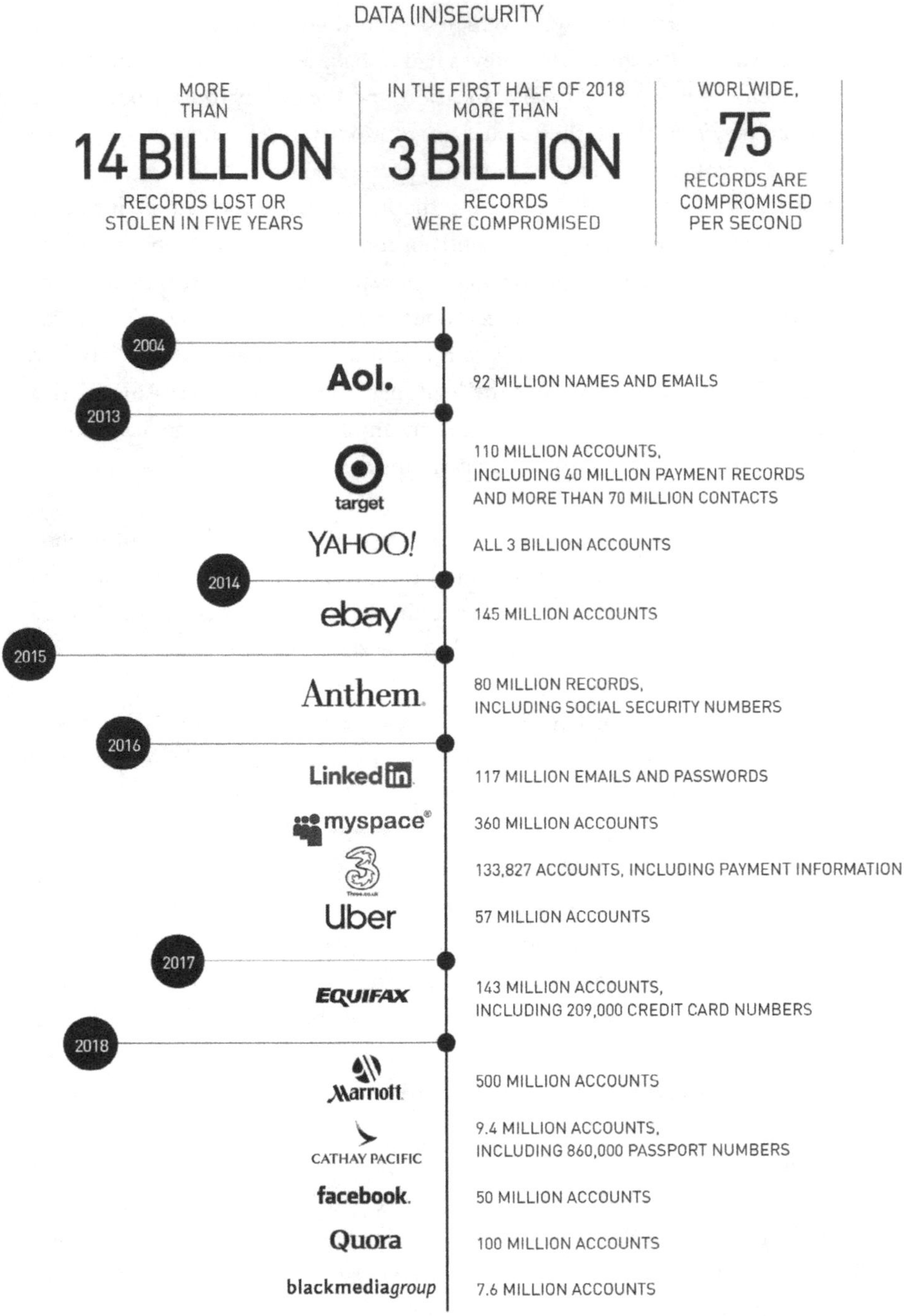

DATA (IN)SECURITY

MORE THAN **14 BILLION** RECORDS LOST OR STOLEN IN FIVE YEARS	IN THE FIRST HALF OF 2018 MORE THAN **3 BILLION** RECORDS WERE COMPROMISED	WORLWIDE, **75** RECORDS ARE COMPROMISED PER SECOND

2004

Aol. — 92 MILLION NAMES AND EMAILS

2013

target — 110 MILLION ACCOUNTS,
INCLUDING 40 MILLION PAYMENT RECORDS
AND MORE THAN 70 MILLION CONTACTS

YAHOO! — ALL 3 BILLION ACCOUNTS

2014

ebay — 145 MILLION ACCOUNTS

2015

Anthem. — 80 MILLION RECORDS,
INCLUDING SOCIAL SECURITY NUMBERS

2016

Linked — 117 MILLION EMAILS AND PASSWORDS

myspace® — 360 MILLION ACCOUNTS

3 Three.co.uk — 133,827 ACCOUNTS, INCLUDING PAYMENT INFORMATION

Uber — 57 MILLION ACCOUNTS

2017

EQUIFAX — 143 MILLION ACCOUNTS,
INCLUDING 209,000 CREDIT CARD NUMBERS

2018

Marriott — 500 MILLION ACCOUNTS

CATHAY PACIFIC — 9.4 MILLION ACCOUNTS,
INCLUDING 860,000 PASSPORT NUMBERS

facebook. — 50 MILLION ACCOUNTS

Quora — 100 MILLION ACCOUNTS

blackmediagroup — 7.6 MILLION ACCOUNTS

Figure 19.9. Sources: Wikipedia, others

These are only a few of the cases that have been documented—the real list is much longer, and many companies elect not to disclose that they have been hacked to avoid issues with their reputation or lawsuits. According to the Michigan-based Ponemon Institute, which has been running studies on data security since 2002, just one-third of the more than 650 professionals interviewed in 2017 believe their organizations allocate adequate resources to manage the security of their information. And one of the most common types of attacks on businesses is ransomware, typically using an invasion technique called a *Trojan horse*.

The name of this type of hack comes from historical accounts that tell how the ancient Greeks hatched a unique strategy to enter the independent city of Troy. After unsuccessfully laying siege to the city for a decade, with the goal of rescuing Helen (the wife of King Menelaus of Sparta, who had been kidnapped by Paris), the Greeks built a giant wooden horse and hid a select force of men inside it. The rest of the Greek army sailed away, leading the Trojans to believe that the war was won. The Trojans brought what they believed to be a victory trophy inside the walled city. The Greek army returned in the middle of the night, and the soldiers hidden inside the horse opened the doors to the city, destroying it (in 1180 BCE).

In a computer Trojan horse attack, the user receives an email, clicks on a fake link, and their data is rendered inaccessible. The information is held for ransom, usually an amount to be paid in cryptocurrencies to make tracing difficult. In 2016 alone, over 350 million new types of malware (programs with objectives that are harmful to users) were detected, and—even worse—an effective ransomware capable of holding the victim's data hostage could be purchased for around $120 on the dark web (which uses the Internet infrastructure but requires a specific set of tools to be accessed).

But the threat gets even more serious. In May 2017, the world learned of the ransomware *WannaCry*, which rapidly propagated, exploiting a vulnerability in Microsoft Windows that had been detected and published a few months prior by another group of hackers. Microsoft provided the necessary patches before *WannaCry* was unleashed, but many companies that hadn't made the updates or that were using versions of the operating system that were no longer being supported were infected. A few days later, the definitive solution to the attack (which had originated in North Korea) was provided—but not before more than 200,000 computers in approximately 150 countries were infected. The sectors impacted included governments, infrastructure (water, power, sewage, gas, rail, airports, highways, and roads), banks, telecom service providers, car manufacturers, and hospitals.

Simply stated, our cities are vulnerable to attacks in a connected society that is overseen by artificial systems. Such attacks may be politically motivated or simply personal, such as the stunt pulled by Vitek Boden. Disgruntled over not having been hired for a municipal job in Queensland, Australia, Boden hacked the utility company's system in 2000 and caused 800,000 liters (211,000 US gallons) of sewage to be discharged into parks and rivers of Maroochy Shire, around 100 km (60 mi) from Brisbane.

Cyberterrorism is considered a real threat by governments around the world, and many believe that future wars will be fought in this arena. With billions of people and objects connected through the IoT, it is imperative that governments, businesses, and individuals are aware of these threats, and that they are trained to seek out the necessary protection.

A SAFE CAREER

With both the market and technology expanding, one of the careers in highest demand will be that of information-security specialists, who are presently in short supply. According to the 2019 International Information System Security Certification Consortium, or (ISC)^2, *Cybersecurity Workforce Study*, the skills gap in the cybersecurity market was a shortage of more than four million professionals—a 145% gap when compared to the estimated number of cybersecurity professionals in the workforce then. The (ISC)^2 is a nonprofit organization that specializes in training and certifications for cybersecurity professionals. The CSO (Chief Security Officer, who is responsible for companies' cybersecurity) has become one of the most critical positions in the structure of any organization. Since many aspects of an IT structure are often outsourced, for example to telecom companies and Internet providers, MSSPs (managed security service providers) have started to show up on the lists of critical functions as well, and they are responsible for the monitoring and protection of their clients' technology infrastructure.

According to market research firm Cybersecurity Ventures, which has offices in the United States and in Israel, cybersecurity products and services expenditures are set to total around $1 trillion globally between 2017 and 2021. Startups around the world are developing products to address this broad field in fraud prevention, automation, identity management, IIoT (Industrial Internet of Things), data confidentiality, development security, and social engineering, to name just a few sub-sectors. Crunchbase, a company that tracks the venture capital scene, reported that almost $10 billion was invested in cybersecurity startups in 2019, compared to $4.2 billion in 2014.

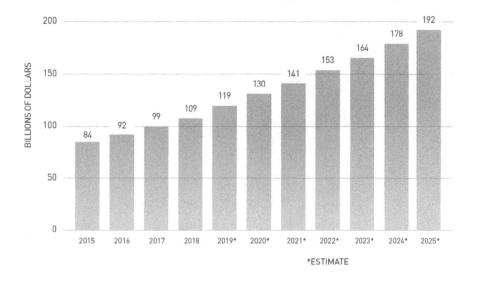

GLOBAL CYBERSECURITY SPEND
(IN BILLIONS OF DOLLARS)

Figure 19.10. Source: *Australia's Cyber Security Sector Competitiveness Plan 2019*, data from Gartner until 2023 extrapolated to 2025 using average growth rates

Two sub-sectors merit a bit more detail.

Development security operations, or DevSecOps, was created to address attacks that occur on the same day as vulnerabilities in the code of a given system are found (a phenomenon known as *zero-day exploit*). To avoid this type of hack, teams specializing in cybersecurity work together with the software engineers during the development stages of a new system.

Social engineering aims to stop users from revealing their passwords in *phishing* schemes. Generally, these schemes are carried out through fake email messages that cause recipients to believe they are interacting with a legitimate company such as a bank, an e-commerce platform, or a large tech company; the messages trick unprepared users into giving their information to hackers. According to the 2018 *Microsoft Security Intelligence Report*, which analyzed nearly half a trillion emails, a 250% increase in instances of phishing had been verified over the previous year.

This type of attack concentrates not on technical vulnerabilities but on the way our brains work: believing, initially, that the sender of the message is legitimate. It is common for cases of phishing to increase after tragedies of a national or international scale—scam emails asking for donations for victims are sent out, aiming at extracting people's financial information (such as credit card numbers) for future use.

PAST FUTURE

OVER THE COURSE OF THIS BOOK, WE HAVE TRAVELED THROUGH
thousands of years of history and met hundreds of people—scientists, scholars, entrepreneurs, statespersons, explorers, and writers—who have contributed in their own unique way to help build a future that belongs to all of us. We have witnessed how the very fabric of civilization is propelled by change, by a natural and ever-growing capacity that we have for asking questions, searching for answers, building hypotheses, making experiments, and developing solutions. The need to change, to create, to invent is at the very core of our nature as a species.

The uniqueness of the Deep Tech Revolution—which is happening here and now, and that will probably continue for a long time—is its breakneck speed. When science becomes technology—by definition, the manifestation of a specific skill—the world changes. Our lives tomorrow are inextricably tied to the research that is happening today, in every single knowledge domain, by fellow humans who are extracting nuggets of truth from nature. What we, as a species, do with this knowledge has never been more important. The very future of humankind depends on it.

The convenience and advances afforded by technology have significantly changed the way we work and interact with each other. We are able to instantly connect with someone on the other side of the world with high-quality images and audio. We ask our devices questions using our native language. We can make payments using our fingerprint or by looking at a biometric sensor. Tests can anticipate and help prevent diseases. We have started to dominate the world of the incredibly small, developing thousands of products and solutions based on nanotechnology. With sensors fitted in living things or inanimate objects, we are able to prepare for, monitor, and optimize industries and processes. The ability to process massive quantities of data using algorithms enables us to forecast the weather, anticipate a client's behavior, create autonomous vehicles, and translate documents in real time. But all this has a cost, and it is essential that we understand the potential consequences of this profound merger of our lives with technology.

The reach of technology—which builds on itself—has reduced the time required between the launch and the adoption of a product or service from decades down to days. Airlines took nearly 70 years to reach 50 million users, cars took a little over 60, the telephone took 50, and TV took 22. But Facebook hit this mark in about three years, WeChat in a year, and the AR game *Pokémon Go* needed only 19 days. This is a testament to the power of networks and the value of Metcalfe's Law, both of which were discussed in Chapter 10.

THAT'S ENTERTAINMENT

The history of how we consume entertainment illustrates different important aspects of our relationship with technology. Videotapes gained popularity in the 1980s, with the VHS format created in 1976 by JVC. Sony's Betamax format, despite having been created a year earlier and having superior image quality, ended up losing market share due to its tapes' shorter recording time and because the company had chosen not to license its technology to other manufacturers. It's not always the company that creates a new market that ends up being the leading force.

Next up came the LaserDisc, based on optical storage technology invented by David Gregg (1923–2001) in 1958, with additional contributions made by James Russell in 1965. It had significantly superior image and sound quality but only took off in the Asian market. Since each side only stored between 30 and 60 minutes of content, the consumer had to turn the disc over manually. In addition, the prices of both the disc player and the movies were too high. At the end of the day, consumers often choose convenience and price over quality.

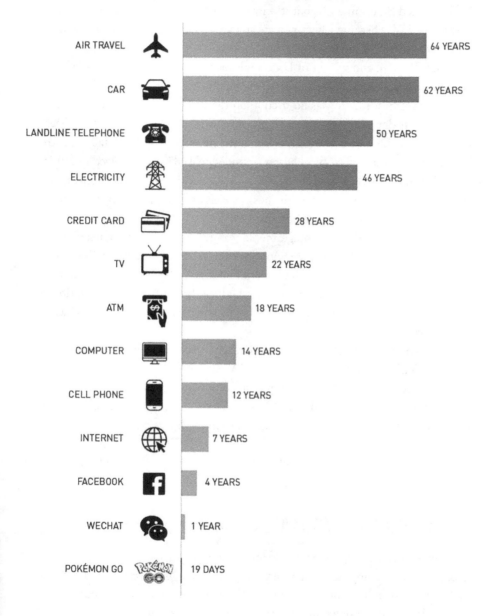

TIME TO REACH 50 MILLION USERS

AIR TRAVEL		64 YEARS
CAR		62 YEARS
LANDLINE TELEPHONE		50 YEARS
ELECTRICITY		46 YEARS
CREDIT CARD		28 YEARS
TV		22 YEARS
ATM		18 YEARS
COMPUTER		14 YEARS
CELL PHONE		12 YEARS
INTERNET		7 YEARS
FACEBOOK		4 YEARS
WECHAT		1 YEAR
POKÉMON GO		19 DAYS

Figure 20.1. Sources: World Economic Forum, others

LaserDisc technology led to the Digital Versatile Disc in 1995, which could store both movies and software. The format was introduced in Japan in 1996 and in the United States in 1997, but it only started to become popular in the early 2000s. The DVD became a huge success at an affordable price and with the same dimensions as a CD (compact disc, used in the recording industry and launched in 1982). The size and weight of DVDs meant that mail-based distribution was economically feasible and led to the 1998 launch of the first online DVD rental store: Netflix. Since there was no friction on the surface of a DVD when it was being read, its durability was considerably greater than a VHS tape. In June 2003, the number of DVD movie rentals eclipsed the number of VHS rentals—a market that was dominated by Blockbuster (founded in 1985 and dissolved in 2013).

Often, the first version of an idea (the LaserDisc) is not the best implementation of a new technology, and business models that do not adapt to the new trends taken up by the public tend not to survive.

The next generation of storage media was the Blu-ray for high-definition TVs. Between 2006 and 2008, we witnessed a new format war, led by Sony with the Blu-ray and Toshiba with the high-definition (HD) DVD. In addition to an aggressive marketing campaign, Sony had the broad base of PlayStation 3 consoles, which could play Blu-ray movies, so there was no need to purchase a new device.

With the popularization and associated convenience of streaming services such as Netflix, Amazon Prime, Apple TV, and Disney+, the rate of use of physical media is likely to keep dropping. Improvements in image and sound standards (such as 4K for video and Dolby Atmos for audio) require networks with higher speeds (which are gradually becoming more accessible), creating space for formats such as Ultra HD Blu-ray and HVD (Holographic Versatile Disc). Once again, we go back to Heraclitus (see Chapter 14): everything flows.

FROM STREAM TO MAINSTREAM

Streaming (the continuous transmission of content to the consumer, with no need for downloading) is omnipresent today. It leverages an extensive telecommunications structure and has increased the speed of transmission and receipt of data. It uses more powerful processors, has prompted the inclusion of operating systems in TVs, and driven the diffusion of portable electronics. In a good example of our past being our future, the history of streaming began in the 1920s, with Major General George Squier (1865–1934).

Figure 20.2. Inventor and holder of a PhD from Johns Hopkins University, Major General George Squier. Source: United States Army Images

Squier was a career officer in the US Army who worked extensively with radio waves and electricity, eventually obtaining a patent for a system that could send out signals over the electricity grid. Radios were still expensive at the time, so his technology generated interest. After successful testing, in which residents of Staten Island were able to hear music that was transmitted over the power lines at their homes, without the use of a radio, the former North American Company (a holding company that controlled several utility companies) acquired the patent rights and formed a company called Wired Radio, Inc. Decades before anything remotely similar would become part of our daily lives, Wired Radio charged for a music streaming service on household electric bills: Subscribers leased a compact receiving instrument and simply plugged it in to any light socket.

But as we know, technological evolution is unrelenting, and just over 10 years later the popularity of radio exploded, causing its price to fall sharply. So Squier changed the focus of his business, setting residential clients aside for commercial ones. In 1934, he changed his company's name to Muzak, inspired by Kodak. And like the photography company, Muzak filed for bankruptcy (in 2012). The term became synonymous with *elevator music*, although this wasn't a market the company had been aiming for.

Starting in the latter half of the 1990s, various initiatives began taking advantage of the concept of Internet streaming since personal computers were gaining the necessary processing capabilities and, at the same time, the speed of household networks was increasing. Microsoft's ActiveMovie (1995), Apple's QuickTime 4 (1999), and Adobe's Adobe Flash (2002) were a few of the milestones in the evolution of this technology, which became a cornerstone for businesses such as Netflix, Spotify, and YouTube.

Netflix started as an online DVD rental service sending DVDs to customers by mail, and between 1998 and 2007 that was its business model. But since it had been born in the digital world, the idea of offering online movies had been a company vision for a long while. They even designed a household device for subscribers that would download the desired movies, to be watched the next day. That idea was shelved, and the model based on streaming was launched instead—in 2007—initially for computers and then gradually for video game consoles, TVs connected to the Internet, and portable devices. By April 2020, the company was offering its services in every country but mainland China, Syria, North Korea, and Crimea, with over 193 million subscribers and a market value of over $175 billion.

The rapid growth of Netflix was due to massive investments in content and commitments worth billions of dollars and has generated a number of initiatives—both in content creation and distribution and in the development of streaming devices—involving players such as Disney and Fox, AT&T and Time Warner, Comcast, Amazon, Apple, and Google. The public's preference for the convenience of streaming and the available content choices are shaping the distribution of movies, shows, concerts, and documentaries. Similarly, the change in how we listen to music—initially on vinyl, then on CD, then from MP3 files, and now via streaming—has also created opportunities for new services and companies. In yet another example of the power of adaptable business models and technology, the drop in CD sales was accompanied by an increase in the revenue earned by recording companies and artists from the royalties paid by the streaming services themselves.

Consumers want to be able to watch or listen to whatever they want, whenever and wherever they like: According to the Digital Entertainment Group, DVD and Blu-ray sales in the United States fell by $6.1 billion in 2015 (a 12% drop over 2014, which in turn saw an 11% drop over 2013). That same year, the revenue earned from digital media grew by nearly 20%. The year 2017 was the first (but not the last) that revenue from streaming exceeded revenue from traditional formats, and services such as Spotify, Apple Music, and Amazon Music generated $6.6 billion (a greater than 40% increase over 2016), representing nearly 40% of the global recorded music industry, while CD sales accounted for 30%.

In February 2005, three former PayPal employees launched an Internet company whose sole objective was to create a simple and convenient method

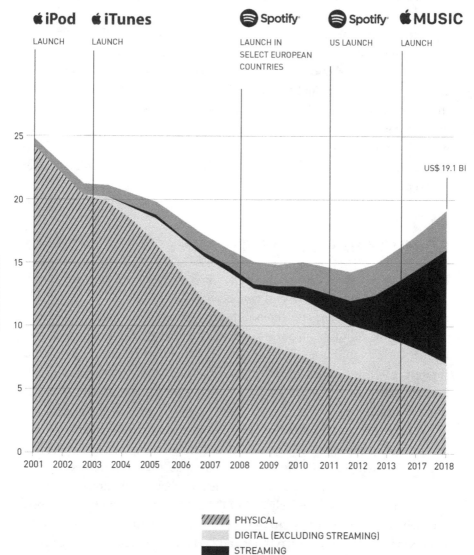

Figure 20.3. Sources: Statista (creative commons), others

for sharing videos. Just 21 months later, in November of 2006, Google acquired YouTube for $1.65 billion. By May 2018, YouTube was the second-most popular website in the world, behind only Google's own website.

YouTube's catalog—on almost any topic from music to religion, sports to cartoons, tutorials to interviews, curiosities to news—is growing at an astonishing rate. In 2015, the company stated it had received around 400 hours of video each minute that year, and this number keeps growing. In 2017, on its official blog, it announced that over one billion hours of video were viewed every day around the world (it would take a single person over 100,000 years to watch this amount of content). TV seems to be the market that has been affected the most. According to a study commissioned by Google and carried out by market research company Nielsen in late 2017, 90% of 18 to 49-year-olds in the United States watched YouTube, and more than half of them were either light viewers or did not subscribe to TV at all.

SO MUCH DATA, SO LITTLE TIME

For most of the world, access to information has never been so easy. Via computers, smartphones, tablets, and connected TVs, we have all the knowledge accumulated by humans over thousands of years at our disposal. We can learn, search, teach, consult, compare, develop, and reason. Programs from some of the world's top universities can be pursued online. People with common interests can form communities to share their opinions, exchange ideas, and arrange meet-ups. Books are available digitally, for instant consumption with our eyes or our ears.

Where, then, are the signs of evolution in our society? Why does it seem we are not moving toward a pluralistic, open, and scientifically and philosophically based society, but rather toward one that is governed by fads, demagoguery, simplistic answers to complex problems, prejudice, and discord?

The answer may be precisely one of the side effects caused by the unprecedented convenience that surrounds us. For the first time in history, we are connected in real time, to everything and everyone; we are informed of events such as earthquakes, shootings, landslides, explosions, and deaths almost instantaneously with the event itself. We check our phones as soon as we wake up to see the latest developments in the world, especially on our social networks. We take one last look at the news before we go to sleep, and between these two moments—waking up and falling asleep—we look at our screens an average of 96 times, according to a late 2019 research study sponsored by global tech care company Asurion.

Our dependence on this connection—and its incessant flow of news, updates, and information—potentially creates a sense of powerlessness, with significant consequences: We spend so much time trying to get to know and understand what is happening that there is little time to actually do something about the very facts we are digesting. But according to a report by the Pew Research Center, in 2016 only 20% of Americans felt overloaded by information compared to 27% a decade before—a surprisingly low number considering the volume of data that hits us during every waking hour, hurting productivity, social ties, and focus.

It takes a deliberate effort to stay disconnected, to remain immune to this non-stop bombardment of data. It took electricity 46 years to reach 50 million people, but the Internet did the same thing in just seven—and we now expect the Internet to be as ubiquitous as electricity, and we don't usually contemplate being off the information grid. The electrification of the world was also accompanied by several of the themes we've seen throughout this book: innovation shrouded in disputes, conflicting egos, and clashes based on marketing and not on technology.

A SHOCKING REALITY

The American inventor Thomas Edison (1847–1931) is one of the figures we think of most in terms of the expansion of electrical systems in the modern world, which effectively made the Second Industrial Revolution possible. With more than 1,000 patents to his name, Edison created a working method that involved a great number of researchers and assistants. His most famous patent, for the first commercially viable incandescent lamp, ended up being disputed in the courts: It was granted in 1880, revoked in 1883 because of the work by William E. Sawyer (1850–1883), and awarded back to Edison in 1886.

In an analogy to the battles between Betamax and VHS and between Blu-ray and HD DVD, the world witnessed a "current war" during the birth of our electrified world. On one side was the direct current (DC) promoted by the Edison Electric Light Company (the precursor of General Electric), and on the other was alternating current (AC), advanced by the Westinghouse Electric Corporation (which acquired the TV and radio network CBS in 1995 and in 1997 became the CBS Corporation), founded by George Westinghouse (1846–1914) in 1886. After an intense marketing battle, involving everything from the electrocution of animals (to demonstrate the dangers of AC) to hiring expensive lawyers to prevent someone sentenced to the electric chair from being executed using Westinghouse's technology, the practical and economic advantages of AC—invented by Nikola Tesla (1856–1943)—ended up winning out.

Figure 20.4. American inventor Thomas Edison (1847–1931). Source: United States Library of Congress's Prints and Photographs division, digital ID cph.3c05139

But the more we get used to a certain technology, the more dependent we become on it. Starting in the latter third of the 1900s, if a power outage occurred, the seven-segment LED display on VCRs would flash "12:00," patiently waiting for someone to reset the time. The procedure for this was relatively simple, but it typically required a quick look-up in the product's manual, and this was something that most people did not have the interest, patience, or time for.

This lack of patience seems to be exacerbated by a culture of—as we have seen—information overload. There's too much to read and to process, not enough time to analyze, and too great a temptation to jump into the easy and simple explanation or justification. This is a recipe for disaster, because as Albert Einstein once said, "everything should be made as simple as possible, but not simpler."

Toward the end of the twentieth century, the world experienced the fear of what could happen in the last minute of December 31, 1999. The so-called Year 2000 problem, known as Y2K, ignited media interest and brought a few countries to high levels of concern (while others simply ignored the issue). The problem consisted basically of the uncertainty of how software would interpret the change from "99" to "00," since the original programmers had represented the year fields with only two digits, rather than four (this memory savings was justifiable at the time when these systems were being planned and implemented). The year 1999 ended and 2000 started with no major hiccups, but the simple fact that we did not know what could happen is very telling: Technology was already more complex and more integrated into our lives than any of us could understand. And the timing of the next potential problem caused by the representation of data on computers is now approaching in our calendars: the early morning of January 19, 2038, at 3:14:07 a.m., Greenwich Mean Time.

The way time is accounted for in different systems—including those based on the Unix operating system, whose development began in Bell Labs, is done by counting the number of seconds elapsed since January 1, 1970. The problem is that the data structure used for this has 32 bits, so the highest number it can represent is just over two billion seconds (or more precisely, 2,147,483,647 seconds) after this date. After this, the time represented will not make sense. Although measures are being taken to get around this bug, the biggest concern is with embedded systems (that is, systems that are incorporated to carry out a specific function within larger, more complex structures) used in fields such as transportation, telecommunications, medical equipment, and electronics.

In the past two decades, this complexity has expanded in all segments, including our homes—the household Internet structure has routers, Wi-Fi access points, and modems. In his 2016 book *Overcomplicated*, American author Samuel Arbesman states that the technological advances that have made our lives more convenient have also created an incomprehensible and unpredictable infrastructure. We are victims of the complexity that is needed to make the use of these technologies as simple as possible.

SIX-TO-ONE ODDS

"I know that I know nothing" is one of the most famous sayings (but was likely never uttered) attributed to the Greek philosopher Socrates (470–399

BCE). This has never been more true, and even the least observant among us will realize the depth and breadth of any single topic search on the web. Any hope we may have had of gaining a broad and deep level of knowledge on multiple topics is dashed by the immediate realization that this is simply no longer possible.

Our capacity for understanding the world is directly related to the tools we possess. As we have evolved as a species, broadening our capacity to listen, see, deduce, reason, extrapolate, and explain, we have broadened the horizons of the possible and unlocked mysteries that had once been unexplainable. We have managed to domesticate animals, decode natural cycles, and harvest all different sorts of crops. We have created language, both oral and written; we have perpetuated knowledge; and we understand the most fundamental laws that govern the universe.

We have discovered the enormity of the cosmos and the immense empty spaces that exist between the galaxies. We have disassembled and reassembled the atom, and we've manipulated the genetic code that is behind all living things. But now we are failing—precisely when scientific discoveries and the most powerful inventions of history are in our hands—at the tasks that are critical to our survival. Eliminating waste. Controlling climate change. Differentiating human from artificial. Separating truth from lies.

Our built-in decision-making system is not able to handle multiple variables at once. Looking for an empty spot in a crowded parking lot is not exactly fun, but it has a simple solution: As soon as a spot is located, all we have to do is park. But drive around a lot with dozens of spots available, and the decision becomes much harder. We waver in the face of so many possibilities. Now, our technology-powered world offers us dozens of available spots, all day long. We are bombarded with information, and we're preconditioned to believe everything that comes across our screens. The days when the news was selected and proven based on the work of a small number of institutions with a high degree of credibility are gone.

How can we tell real news from fake news? Organized networks of disinformation pollute the virtual space with gossip, intrigue, conspiracy theories, and the so-called deepfake videos. Their objective is to sow chaos, fear, doubt, and uncertainty and to stop us from being able to analyze the facts of a story and its developments. The same tools that inform us also disinform us; the ones that present us with facts also trick us with rumors. We will need more technology to solve this problem, and this solution will be followed by more technology to generate more disinformation in a cycle similar to the dynamic between guardians of a fortress and potential raiders.

In a 2018 paper by MIT scholars titled "The Spread of True and False News Online," the behavior of Twitter users toward a data set of 126,000 news articles, fact-checked between 2006 and 2017, was analyzed. Truth rarely reached more than 1,000 people, whereas some false stories—especially political news—got to more than 10,000 readers. And it is a fact that lies spread faster than truth: in this study, by a factor of six. In other words, it took true stories six times longer to reach the same number of users than the dreaded fake news, not due to bots, but due to humans.

The excess of data generated by everything and everyone has another negative aspect: It is possible to build arguments on practically any topic and to assemble a data set that, at first glance, supports its thesis. This phenomenon is so complex that even highly qualified researchers end up falling victim to it. In a 2005 article published in the *Public Library of Science Medicine (PLoS Medicine)* by Greek-American physician and researcher John Ioannidis, titled "Why Most Published Research Findings Are False," the argument presented mentions how statistical tests are being used to validate contentions that should be challenged, or at least questioned.

This fundamental issue—fake news against the truth, disinformation against reality, deepfakes against facts—may only be addressed by a meticulous analysis of the mechanisms that permeate the whole process: the birth of a lie, its spread across users in one or multiple social networks, the algorithms managed by those very social networks that serve their users with disinformation, and the behavior that makes people much more likely to spread a falsehood than a true statement. This difficulty, at the very intersection of business, science, and technology, is one of our key challenges going forward as a society.

WRONG QUESTIONS

The challenge of a world with an excess of data is captured in the term Eroom's Law, which is the opposite of Moore's Law, which we discussed in Chapter 4. Eroom is Moore spelled backward, and it arose from the observation that the discovery of new medications is becoming slower and more expensive despite significant advances in biotechnology and computing. This slowdown was documented in the 2012 article "Diagnosing the decline in pharmaceutical R&D

efficiency," published in *Nature Reviews Drug Discovery* and written by Jack Scannell, Alex Blanckley, Helen Boldon, and Brian Warrington (all of them connected to the world of investments). According to the article, this slowdown occurs for reasons like the existence of drugs well suited to the task, which thus inhibits the development of substitutes, the more conservative approach taken by regulators, and the use of basic research and development methods that are inadequate for dealing with complex diseases.

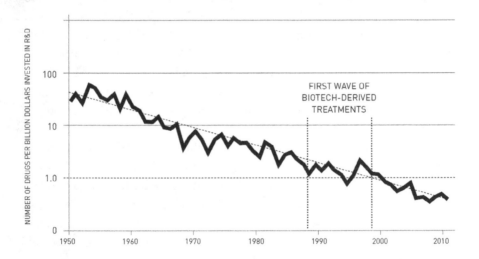

EFFICIENCY OF NEW DRUG
RESEARCH & DEVELOPMENT
(ADJUSTED FOR INFLATION)

Figure 20.5. Eroom's law and drug discovery: slower and more expensive, instead of faster and cheaper. Source: J.W. Scannell, A. Blanckley, H. Boldon, B. Warrington

A world ruled by big data runs the risk of becoming a world without explanations. As we have seen, this technology uses the correlation among all data—as unintuitive as these links may seem—to generate answers, without necessarily understanding the reasons behind the solution returned. In his 2018 book, *New Dark Age: Technology and the End of the Future*, English writer and artist James Bridle articulates that "the big data fallacy is the logical outcome of scientific reductionism: the belief that complex systems can be understood by dismantling them into their constituent pieces and studying each in isolation."

The immense data sets we produce include valuable information on what is being consumed online. Which videos are watched? Which songs are listened to? Who has more followers? Which websites are most visited?

The importance of keeping precise metrics on the Internet was quickly realized—which is not surprising, considering the tendency engineers have for quantifying everything possible. In light of this, computing engineer Brewster Kahle, along with Bruce Gilliat, founded Alexa Internet in 1996. The inspiration for the name Alexa came from the famous Library of Alexandria, likely the most important of ancient times, which housed virtually all knowledge accumulated by civilization up to that time. The company's objective was to analyze Internet traffic. Before founding Alexa, Kahle worked on the WAIS (Wide Area Information Server) system, which was similar to what would become HTTP (Hypertext Transfer Protocol)—effectively the language used by websites all around the world. In 2001, he launched the website *Wayback Machine*, which was developed to enable access to old versions of any webpage stored in the Internet Archive (which was founded in 1996).

In 1999, Amazon acquired Alexa Internet, which started producing the Alexa Rank, wherein websites are classified based on an estimate of their popularity. This estimate uses data such as the number of daily visitors and the number of pages viewed, thus providing an idea of what is being accessed on the web. In January 2019, the top 10 slots were occupied by Google, YouTube, Facebook, Baidu, Wikipedia (great news!), Tencent QQ, Taobao, Tmall, Yahoo!, and Amazon (six US-based companies and four Chinese ones).

When looking at the most-watched videos on YouTube, we see that in August 2019 only three of the top 30 were not professional music videos. According to SEO firm Ahrefs (SEO is the acronym for search engine optimization), excluding the brands and words related to pornography, the three most-searched words in May 2019 were *weather*, *maps*, and *translate*.

In other words, the most complex general-use tool in the history of civilization basically works as a social network to watch music videos, share photos, make purchases, watch pornography, and carry out searches which, in turn, lead to other websites (such as Facebook, YouTube, and Amazon) or to apps that have been specifically developed for these purposes (such as weather apps, step-by-step directions on maps, and translation of various languages).

PRESENT FUTURE

American science-fiction writer Philip K. Dick (1928–1982) is known for the film adaptations of many of his works, such as *Blade Runner* (1982), based on *Do Androids Dream of Electric Sheep?* (1968); *Total Recall* (1990 and 2012), based on the short story *We Can Remember It for You Wholesale* (1966); and *Minority Report* (2002), based on the short story *The Minority Report* (1956). The 2017 TV series *Electric Dreams* was also based on his work. Among his favorite topics are issues such as what defines humans, interference by governments who know everything and see everything about the lives of their citizens, and megacorporations with limitless powers.

We have arrived at a moment in our history where these three themes are extremely pertinent when discussing society, values, and principles. We have seen that talking about a distant, remote future doesn't make sense—the future is already part of our present. Citing science-fiction author William Gibson, the future is already here—it's just not very evenly distributed.

Changes happen constantly, before our eyes. We don't notice the growth of our children because we see them every day. But it only takes someone who doesn't see us regularly to say, "They've grown so much!" and we are reminded of the relentless march of time, which takes everything and everyone along with it, without distinction. We are living in a time when the excess of choices, options, and alternatives can leave us paralyzed. The complexity and convenience of the world lived in by a large part of the population causes anxiety and a sense of powerlessness that suppresses ambition, inhibits curiosity, and confounds expectations.

In the 1926 book *The Sun Also Rises*, by Ernest Hemingway (1899–1961), the question "How did you go bankrupt?" is posed to one of the characters. His answer: "Two ways. Gradually, then suddenly." That's exactly how the future is arriving: gradually, then suddenly. Autonomous vehicles, artificial intelligence, the IoT, biotechnology, 3D printing, VR, robotics, nanotechnology—all of them are already here, unevenly distributed though they may be.

The challenges posed by the present future demand that we are always mindful and do not succumb to the temptation of likes and followers. The world is round. Vaccines save lives. Global warming is real. These are truths that have been scientifically established and that do not depend on subjective judgments or personal opinions. But they are facts that, in a return to a medieval past, have started being debated again, for no rational reason. It is essential that we maintain a critical perspective and keep an eye on the developments of the exponential changes underway (which are sure to continue at their rapid pace, unevenly distributing the future itself). These are essential abilities for us humans, who are surrounded by our own works—the fruit of hundreds of generations of creators, dreamers, and inventors.

The future is not only present. It is a present. Use it wisely.

ACKNOWLEDGMENTS

"IF THERE'S A BOOK THAT YOU WANT TO READ, BUT IT HASN'T been written yet, then you must write it," said Toni Morrison (1931-2019), the great American author who won the 1993 Nobel Prize in Literature. This is the reason why I wrote *Present Future*.

I would like to thank the team that helped me bring the book you have just read to your hands. Dakin Sloss, founder and general partner at Prime Movers Lab, introduced me to Tanya Hall, CEO at Greenleaf Book Group. She immediately bought into the premise of the book, and Patrick Hainault at Fast Company agreed (in record time!). Thanks to Sam Alexander, Tiffany Barrientos, Tyler LeBleu, Neil Gonzalez, Scott James, Emily Maulding, Danny Sandoval and Tonya Trybula for applying their expertise into this work.

Thanks to Arminio Fraga and Josh Wolfe, friends and extraordinary investors and scholars for writing the preface for the Brazilian and US editions respectively. To Pedro Cappeletti, a designer of exceptional talent: Thank you for turning cold, hard data into interesting and engaging illustrations. To Andrea Roach, my detail-oriented translator, thanks for making this part of the process so smooth.

And, of course, a very special thanks to my editors, Sally Garland and Dr. Heather Stettler, who made the manuscript you just read so much better and enjoyable with their questions, comments, and suggestions. Any errors or omissions are my own responsibility.

To my dear friend Mario Bomfim, thank you for the more than 30 years of friendship. To my sister Catherine and my brothers-in-law Paulo and

Luiz Antônio, thank you for all the encouragement you have given me along this journey.

To my wife, Michelle, and my children, Dan and Ingrid, thank you for putting up with the long evenings (and nights, and early mornings) while I was working on organizing this book in the best possible way—and thank you for cheering me on. To my father, Armand, who I hope will one day write his own book, all my love. If there is a heaven, my mother is there, and she has already read the book 18 times in a row.

But my biggest thanks go to my youngest sister, Isabelle, without whom this book would not have become a reality. She was an unconditional supporter of this project from day one, with incomparable attention to detail. This book would never have been possible without her.

GUY PERELMUTER
FEBRUARY 2021

INDEX

BIBLIOGRAPHY

PREFACE

1. Bracetti, Alex. "The 25 Craziest Things Ever Said by Tech CEOs," *Complex,* January 14, 2012. https://www.complex.com/pop-culture/2013/01/the-25-craziest-things-ever-said-by-tech-ceos/.

2. O'Toole, Garson. "People Tend to Overestimate What Can Be Done in One Year and to Underestimate What Can Be Done in Five or Ten Years," Quote Investigator. January 3, 2019. https://quoteinvestigator.com/2019/01/03/estimate/.

3. Wikipedia, s.v. "J. C. R. Licklider." https://en.wikipedia.org/wiki/J._C._R._Licklider.

4. MacMullen, W. John. "Bob Gets His Just Desserts . . . " Ibiblio, April 11, 1997. https://www.ibiblio.org/pjones/ils310/msg00259.html.

5. Marshall, Michael. "10 Impossibilities Conquered by Science," *New Scientist,* April 3, 2008, Space. https://www.newscientist.com/article/dn13556-10-impossibilities-conquered-by-science/.

6. McKinley, Joe. "13 Predictions About the Future That Were Dead Wrong," *Reader's Digest, n.d.* Culture. https://www.rd.com/culture/predictions-that-were-wrong/.

7. Pogue, David. "Use It Better: The Worst Tech Predictions of All Time," *Scientific American,* January 18, 2012. https://www.scientificamerican.com/article/pogue-all-time-worst-tech-predictions/.

8. Torkington, Simon. "10 Predictions for the Future that Got It Wildly Wrong," *World Economic Forum,* October 13, 2016. Formative Content. https://www.weforum.org/agenda/2016/10/10-predictions-for-the-future-that-got-it-wildly-wrong/.

9. *Telegraph.* "Worst Tech Predictions of All Time." June 29, 2016. https://www. telegraph.co.uk/technology/0/worst-tech-predictions-of-all-time/.

CHAPTER 1—THE WORLD POWERED BY TECHNOLOGY

10. Morris, Ian. *Why the West Rules—for Now: The Patterns of History, and What They Reveal About the Future.* Picador, 2011.

11. Morris, Ian. *The Measure of Civilization: How Social Development Decides the Fate of Nations.* Princeton University Press, 2014.

12. Schwab, Klaus. "The Fourth Industrial Revolution: What It Means, How to Respond," World Economic Forum, January 14, 2016, https://www.weforum. org/agenda/2016/01/the-fourth-industrial-revolution-what-it-means-and-how-to-respond.

CHAPTER 2—AUTONOMOUS VEHICLES

13. Barber, Megan. "Before Tesla: Why Everyone Wanted an Electric Car in 1905," *Curbed,* September 22, 2017. https://www.curbed.com/2017/9/22/16346892/ electric-car-history-fritchle.

14. Wikipedia, s.v. "Code of Hammurabi." https://en.wikipedia.org/wiki/Code_of_ Hammurabi.

15. Domm, Patti. "Electric Vehicles: The Little Industry That Could Take a Bite out of Oil Demand," *CNBC,* February 28, 2018. https://www.cnbc. com/2018/02/28/soon-electric-vehicles-could-cause-an-oil-crisis-.html.

16. *Historic Wings.* "George the Autopilot," August 30, 2012. http://fly. historicwings.com/2012/08/george-the-autopilot/.

17. Wikipedia, s.v. "Global Positioning System." https://en.wikipedia.org/wiki/ Global_Positioning_System.

18. "How Does GPS work?" *Physics.org,* n.d. http://www.physics.org/article-questions.asp?id=55.

19. IEA. "Global EV Outlook 2020: Entering the decade of electric drive?" June 2020. https://www.iea.org/reports/global-ev-outlook-2020.

20. "National Motor Vehicle Crash Causation Survey," US Department of Transportation, July 2008. https://crashstats.nhtsa.dot.gov/Api/Public/ ViewPublication/811059.

21. Randall, Tom. "Here's How Electric Cars Will Cause the Next Oil Crisis," *Bloomberg,* February 25, 2016. http://www.bloomberg.com/features/2016-ev-oil-crisis/.

22. Richardson, Jake. "38% of American Cars Were Electric in 1900," *CleanTechnica*, February 25, 2018. https://cleantechnica.com/2018/02/25/38-percent-american-cars-electric-1900/.

23. Scheck, William. "Lawrence Sperry: Genius on Autopilot," *Aviation History* magazine, November 2004, republished on *HistoryNet*. https://www.historynet.com/lawrence-sperry-autopilot-inventor-and-aviation-innovator.htm.

24. Strohl, Dan. "Ford, Edison and the Cheap EV That Almost Was," *Wired,* June 18, 2010, Gear. https://www.wired.com/2010/06/henry-ford-thomas-edison-ev/.

25. Wikipedia, s.v. "The Trolley Problem." https://en.wikipedia.org/wiki/Trolley_problem.

CHAPTER 3—THE FUTURE OF JOBS

26. Allen, Katie. "Technology Has Created More Jobs than It Has Destroyed, Says 140 Years of Data," *Guardian*, August 18, 2015. https://www.theguardian.com/business/2015/aug/17/technology-created-more-jobs-than-destroyed-140-years-data-census?CMP=share_btn_twZ.

27. Wikipedia, s.v. "Basic income around the world." https://en.wikipedia.org/wiki/Basic_income_around_the_world.

28. Brynjolfsson, Erik, and Andrew McAfee. *The Second Machine Age: Work, Progress, and Prosperity in a Time of Brilliant Technologies*, W.W. Norton & Company, 2016.

29. Ford, Martin. *Rise of the Robots: Technology and the Threat of a Jobless Future.* Basic Books, 2016.

30. Frey, Carl Benedikt, and Michael A. Osborne. "The Future of Employment: How Susceptible Are Jobs to Computerisation?" University of Oxford website, September 17, 2013. http://www.oxfordmartin.ox.ac.uk/downloads/academic/The_Future_of_Employment.pdf.

31. Graetz, Georg, and Guy Michaels. "Robots at Work," Centre for Economic Policy Research, March 2015. http://cepr.org/active/publications/discussion_papers/dp.php?dpno=10477.

32. Hilsenrath, Jon, and Bob Davis. "America's Dazzling Tech Boom Has a Downside: Not Enough Jobs," *Wall Street Journal,* October 12, 2016. http://www.wsj.com/articles/americas-dazzling-tech-boom-has-a-downside-not-enough-jobs-1476282355.

33. Hamari, Juho, Mimmi Sjöklint, and Antti Ukkonen, "The Sharing Economy: Why People Participate in Collaborative Consumption," *Journal of the Association for Information Science and Technology* 67(9): 2047–2059, September 2016. https://www.researchgate.net/publication/255698095_The_Sharing_Economy_Why_People_Participate_in_Collaborative_Consumption.

34. "History of Basic Income," Basic Income Earth Network, n.d. http://basicincome.org/basic-income/history/.

35. Konrad, Alex. "From Communism to Coding: How Daniel Dines of $7 Billion UiPath Became the First Bot Billionaire," *Forbes,* September 11, 2019. https://www.forbes.com/sites/alexkonrad/2019/09/11/from-communism-to-coding-how--daniel-dines-of-7-billion-uipath-became-the-first-bot-billionaire/#173fde00206e.

36. Muro, Mark, and Scott Andes. "Robots Seem to Be Improving Productivity, Not Costing Jobs," *Harvard Business Review,* June 16, 2015. https://hbr.org/2015/06/robots-seem-to-be-improving-productivity-not-costing-jobs.

37. Rotman, David. "How Technology Is Destroying Jobs," *MIT Technology Review,* June 12, 2013, Business Impact. https://www.technologyreview.com/s/515926/how-technology-is-destroying-jobs/.

38. Wikipedia, s.v. "Technological unemployment." https://en.wikipedia.org/wiki/Technological_unemployment.

39. Thompson, Derek. "What Jobs Will the Robots Take?" *The Atlantic,* January 23, 2014, Business. http://www.theatlantic.com/business/archive/2014/01/what-jobs-will-the-robots-take/283239/.

CHAPTER 4—ARTIFICIAL INTELLIGENCE

40. Wikipedia, s.v. "Ada Lovelace." https://en.wikipedia.org/wiki/Ada_Lovelace.

41. Caliskan-Islam, Aylin, Joanna J. Bryson, and Arvind Narayanan. "Semantics Derived Automatically from Language Corpora Necessarily Contain Human Biases," *Random Walker,* August 25, 2016. http://randomwalker.info/publications/language-bias.pdf.

42. Copeland, Jack. "Alan Turing: The Codebreaker Who Saved 'Millions of Lives'," *BBC.* n.p., June 19, 2012, Technology. https://www.bbc.com/news/technology-18419691.

43. Debrule, Sam. "The Non-Technical Guide to Machine Learning & Artificial Intelligence," *Machine Learnings,* November 16, 2016. https://machinelearnings.co/a-humans-guide-to-machine-learning-e179f43b67a0.

44. Descartes, René. *Discourse on Method.* SMK Books, 2009. First published 1637.

45. Diakopoulos, Nicholas. "Accountability in Algorithmic Decision Making," Nick Diakopoulos's website, February 2016. http://www.nickdiakopoulos.com/wp-content/uploads/2016/03/Accountability-in-algorithmic-decision-making-Final.pdf.

46. Groskopf, Christopher. "When Computers Learn Human Languages, They Also Learn Human Prejudices," *Quartz,* August 29, 2016. https://qz.com/768567/when-computers-learn-human-languages-they-also-learn-human-prejudices/.

47. Wikipedia, s.v. "History of artificial intelligence." https://en.wikipedia.org/wiki/History_of_artificial_intelligence.

48. Hodges, Andrew. *Alan Turing: The Enigma: The Book That Inspired the Film The Imitation Game*. Princeton University Press, 2014. First published 1983.

49. Hof, Robert D. "Deep Learning," *MIT Technology Review*, April 23, 2013, Intelligent Machines. https://www.technologyreview.com/s/513696/deep-learning/.

50. Wikipedia, s.v. "Iliad." https://en.wikipedia.org/wiki/Iliad.

51. International Labor Organization. ILOSTAT. https://ilostat.ilo.org/.

52. Levesque, Hector J., Ernest Davis, and Leora Morgenstern. "The Winograd Schema Challenge," Proceedings of the Thirteenth International Conference on Principles of Knowledge Representation and Reasoning, pp. 552–561, 2012. https://www.aaai.org/ocs/index.php/KR/KR12/paper/download/4492/4924.

53. Manyika, James, Susan Lund, Michael Chui, Jacques Bughin, Jonathan Woetzel, Parul Batra, Ryan Ko, and Saurabh Sanghvi. "Jobs Lost, Jobs Gained: Workforce Transitions in a Time of Automation," McKinsey & Company, December 2017. https://www.mckinsey.com/~/media/mckinsey/featured%20insights/future%20of%20organizations/what%20the%20future%20of%20work%20will%20mean%20for%20jobs%20skills%20and%20wages/mgi-jobs-lost-jobs-gained-report-december-6-2017.ashx.

54. McCarthy, J., M.L. Minsky, N. Rochester, and C.E. Shannon. "A Proposal for the Dartmouth Summer Research Project on Artificial Intelligence," Stanford University, August 31, 1955. http://www-formal.stanford.edu/jmc/history/dartmouth/dartmouth.html.

55. McCorduck, Pamela. *Machines Who Think: A Personal Inquiry into the History and Prospects of Artificial Intelligence*. A K Peters/CRC Press, 2004.

56. Press, Gil. "Forrester Predicts Investment in Artificial Intelligence Will Grow 300% in 2017," *Forbes*, November 1, 2016. https://www.forbes.com/sites/gilpress/2016/11/01/forrester-predicts-investment-in-artificial-intelligence-will-grow-300-in-2017/#67670ff55509.

57. Shelley, Mary. *Frankenstein*. Dover Publications, 1994. First published 1818.

58. Wikipedia, s.v. "Stuart J. Russell," n.p., n.d. https://en.wikipedia.org/wiki/Stuart_J._Russell.

59. Tarantola, Andrew. "How to Build Turing's Universal Machine," *Gizmodo*, March 15, 2012. https://gizmodo.com/5891399/how-to-build-turings-universal-machine.

60. "The 2016 AI Recap: Startups See Record High in Deals and Funding." CB Insights website, January 19, 2017. https://www.cbinsights.com/blog/artificial-intelligence-startup-funding/.

61. Turing, A.M. "On Computable Numbers, with an Application to the Entscheidungsproblem," *Proceedings of the London Mathematical Society*, Volume s2–42, Issue 1, 1937, pp. 230–265. https://londmathsoc.onlinelibrary.wiley.com/doi/abs/10.1112/plms/s2-42.1.230.

62. Zilis, Shivon, and James Cham. "The Current State of Machine Intelligence 3.0," Shivon Zilis's website, n.d. http://www.shivonzilis.com/machineintelligence.

CHAPTER 5—THE INTERNET OF THINGS AND SMART CITIES

63. Buntz, Brian. "The World's 5 Smartest Cities," *IoT World Today*, May 18, 2016. http://www.ioti.com/smart-cities/world-s-5-smartest-cities.

64. Department of Economic and Social Affairs, Population Division. "World Urbanization Prospects: The 2018 Revision," United Nations website, n.d. https://population.un.org/wup/Publications/Files/WUP2018-Report.pdf.

65. Dutcher, Jennifer. "Data Size Matters [Infographic]," Berkeley School of Information website, November 6, 2013. https://datascience.berkeley.edu/big-data-infographic/.

66. "Ericsson Mobility Report on the Pulse of the Networked Society," Ericsson website, November 2016. https://www.ericsson.com/assets/local/mobility-report/documents/2016/ericsson-mobility-report-november-2016.pdf.

67. Evans, Dave. "The Internet of Things," Cisco website, April 2011. http://www.cisco.com/c/dam/en_us/about/ac79/docs/innov/IoT_IBSG_0411FINAL.pdf.

68. "Global Agriculture Towards 2050," Food and Agriculture Organization of the United Nations website, Rome, October 12–13, 2009. http://www.fao.org/fileadmin/templates/wsfs/docs/Issues_papers/HLEF2050_Global_Agriculture.pdf.

69. "Global Health Expenditure Database," World Health Organization website. http://apps.who.int/nha/database.

70. "IESE Cities in Motion Index 2018," IESE Business School—University of Navarra website. https://blog.iese.edu/cities-challenges-and-management/2018/05/23/iese-cities-in-motion-index-2018/.

71. Ireland, Tom. "The artificial meat factory—the science of your synthetic supper," *Science Focus*, May 23, 2019. https://www.sciencefocus.com/future-technology/the-artificial-meat-factory-the-science-of-your-synthetic-supper/.

72. Johnson, R. Colin. "Roadmap to Trillion Sensors Forks," *EE Times*, October 12, 2015. Internet of Things Design Line. http://www.eetimes.com/document.asp?doc_id=1328466.

73. "Largest Cities in the World," City Mayors Statistics website, March 2018. http://www.citymayors.com/statistics/largest-cities-population-125.html.

74. Leiner, Barry M., Vinton G. Cerf, David D. Clark, Robert E. Kahn, Leonard Kleinrock, Daniel C. Lynch, Jon Postel, Larry G. Roberts, and Stephen Wolff. "Brief History of the Internet," *Internet Society,* 1997. http://www.internetsociety.org/internet/what-internet/history-internet/brief-history-internet.

75. Lontoh, Sonita. "How Much Is the Internet of Things Worth to the Global Economy?" *World Economic Forum,* April 20, 2016. https://www.weforum.org/agenda/2016/04/how-much-is-the-internet-of-things-worth-to-the-global-economy/.

76. O'Brien, Kevin J. "Talk to Me, One Machine Said to the Other," *New York Times,* July 29, 2012. http://www.nytimes.com/2012/07/30/technology/talk-to-me-one-machine-said-to-the-other.html.

77. Orwell, George. *1984.* Penguin Group, 1983. First published 1949.

78. Population Reference Bureau website. https://www.prb.org/.

79. Simon, Matt. "Lab-Grown Meat Is Coming, Whether You Like It or Not," *Wired,* February 16, 2018. Science. https://www.wired.com/story/lab-grown-meat/.

80. "Singapore Named 'Global Smart City—2016'," Juniper Research website. https://www.juniperresearch.com/press/press-releases/singapore-named-global-smart-city-2016.

81. "Smart Cities—Frost & Sullivan Value Proposition," Frost & Sullivan website, January 2019. https://ww2.frost.com/wp-content/uploads/2019/01/SmartCities.pdf.

82. "The Top 50 Smart Cities in the World 2018," CITI IO website, July 27, 2018. https://www.citi.io/2018/07/27/the-top-50-smart-cities-in-the-world-2018/.

83. World Bank Open Data website. The World Bank. https://data.worldbank.org/.

84. "U.S. MoneyTree Reporting—Explore the Data," PwC and CB Insights. PWC website. https://www.pwc.com/us/en/industries/technology/moneytree/explorer.html#/.

85. Walt, Vivienne. "Is This Tiny European Nation a Preview of Our Tech Future?" *Fortune,* April 7, 2017, Tech Estonia. http://fortune.com/2017/04/27/estonia-digital-life-tech-startups/.

86. Watts, Jake Maxwell, and Newley Purnell. "Singapore Is Taking the 'Smart City' to a Whole New Level," *Wall Street Journal,* April 24, 2016. https://www.wsj.com/articles/singapore-is-taking-the-smart-city-to-a-whole-new-level-1461550026.

87. Wigmore, Ivy. "IPv6 Addresses—How Many Is That in Numbers?" *IT Knowledge Exchange,* January 14, 2009. http://itknowledgeexchange.techtarget.com/whatis/ipv6-addresses-how-many-is-that-in-numbers/.

88. "World Population Trends," United Nations Population Fund website, n.d. http://www.unfpa.org/world-population-trends.

CHAPTER 6—BIOTECHNOLOGY

89. Demetriou, Danielle. "Japan's Elderly Overtake Teenagers in the Shoplifting Stakes," *Telegraph*, July 9, 2013. https://www.telegraph.co.uk/news/worldnews/asia/japan/10167901/Japans-elderly-overtake-teenagers-in-the-shoplifting-stakes.html.

90. Duffy, Maureen. "New Research from Google Labs: Using Machine Learning to Detect Diabetic Eye Disease," VisionAware website, December 9, 2016. http://www.visionaware.org/blog/visionaware-blog/new-research-from-google-labs-using-machine-learning-to-detect-diabetic-eye-disease/12.

91. Cave, Stephen. *Immortality: The Quest to Live Forever and How It Drives Civilization.* Crown, 2012.

92. Cumbers, John. "Synthetic Biology Has Raised $12.4 Billion. Here Are Five Sectors It Will Soon Disrupt," *Forbes,* September 9, 2019. https://www.forbes.com/sites/johncumbers/2019/09/04/synthetic-biology-has-raised-124-billion-here-are-five-sectors-it-will-soon-disrupt/.

93. Easterbrook, Gregg. "What Happens When We All Live to 100?" *The Atlantic,* October 2014. https://www.theatlantic.com/magazine/archive/2014/10/what-happens-when-we-all-live-to-100/379338/.

94. "Focus on Health Spending—OECD Health Statistics 2015," OECD website, July 2015. https://www.oecd.org/health/health-systems/Focus-Health-Spending-2015.pdf.

95. Genome Information by Organism. National Center for Biotechnology Information. https://www.ncbi.nlm.nih.gov/genome/browse/#!/overview/.

96. "Global Health and Aging," National Institute on Aging website, June 2017. https://www.nia.nih.gov/sites/default/files/2017-06/global_health_aging.pdf.

97. Gulshan, Varun, Lily Peng, Marc Coram, Martin C. Stumpe, Derek Wu, Arunachalam Narayanaswamy, Subhashini Venugopalan, Kasumi Widner, Tom Madams, Jorge Cuadros, Ramasamy Kim, Rajiv Raman, Philip C. Nelson, Jessica L. Mega, and Dale R. Webster. "Development and Validation of a Deep Learning Algorithm for Detection of Diabetic Retinopathy in Retinal Fundus Photographs," *Journal of the American Medical Association,* 2016. Vol. 316, Issue 22, pp. 2402–2410. https://jamanetwork.com/journals/jama/fullarticle/2588763.

98. Wikipedia, s.v. "Har Gobind Khorana." https://en.wikipedia.org/wiki/Har_Gobind_Khorana.

99. Wikipedia, s.v. "Hayflick limit." https://en.wikipedia.org/wiki/Hayflick_limit.

100. "Health Care in the Digital Age," Milken Institute website, n.d. http://assets1c.milkeninstitute.org/assets/Events/Conferences/GlobalConference/2015/Slide/MON-Beverly-Hills-Ballroom-1045-SK-Health-Care-in-the-Digital-Age.pdf.

101. "Health Expenditures," National Center for Health Statistics website, n.d. https://www.cdc.gov/nchs/fastats/health-expenditures.htm.

102. Henig, Robin Marantz. *The Monk in the Garden: The Lost and Found Genius of Gregor Mendel, the Father of Genetics.* Mariner Books, 2000.

103. Wikipedia, s.v. "History of genetics." https://en.wikipedia.org/wiki/History_of_genetics.

104. "Human Genome Project FAQ," National Human Genome Research Institute website, n.d. https://www.genome.gov/11006943/human-genome-project-completion-frequently-asked-questions/.

105. Koettl, Johannes. "Boundless Life Expectancy: The Future of Aging Populations," Brookings Institution website, March 23, 2016. https://www.brookings.edu/blog/future-development/2016/03/23/boundless-life-expectancy-the-future-of-aging-populations/.

106. Lander, Eric S. "Brave New Genome," *New England Journal of Medicine,* June 3, 2015. https://www.nejm.org/doi/full/10.1056/NEJMp1506446.

107. Li, Xiao, Jessilyn Dunn, Denis Salins, Gao Zhou, Wenyu Zhou, Sophia Miryam Schüssler-Fiorenza Rose, Dalia Perelman, Elizabeth Colbert, Ryan Runge, Shannon Rego, Ria Sonecha, Somalee Datta, Tracey McLaughlin, and Michael P. Snyder. "Digital Health: Tracking Physiomes and Activity Using Wearable Biosensors Reveals Useful Health-Related Information," *PLOS Biology,* January 12, 2017. http://journals.plos.org/plosbiology/article?id=10.1371/journal.pbio.2001402.

108. Oeppen, Jim, and James W. Vaupel. "Broken Limits of Life Experience," *Science,* Vol. 296, Issue 5570, pp. 1029–1031. May 10, 2002. Policy Forum. http://science.sciencemag.org/content/296/5570/1029.

109. Wikipedia, s.v. "Organ printing." https://en.wikipedia.org/wiki/Organ_printing.

110. Peters, Adele. "3D Printing Living Organs, And Other World-Changing Ideas in Health," *Fast Company,* April 12, 2017. https://www.fastcompany.com/40404565/3-d-printing-living-organs-and-other-world-changing-ideas-in-health.

111. "World Population Prospects 2019," United Nations website, n.d. https://esa.un.org/unpd/wpp/Graphs/DemographicProfiles/.

112. Prentice, Thomson. "Health, History and Hard Choices: Funding Dilemmas in a Fast-Changing World," World Health Organization website, August 2006. http://www.who.int/global_health_histories/seminars/presentation07.pdf.

113. Steger, Isabella. "The Next Big Innovation in Japan's Aging Economy is Flushable Adult Diapers," *Quartz,* January 28, 2019. https://qz.com/1534975/the-next-big-innovation-in-aging-japan-flushable-adult-diapers/.

114. Wikipedia, s.v. "Synthetic biology." https://en.wikipedia.org/wiki/Synthetic_biology.

115. "The Top 10 Causes of Death," World Health Organization website, May 24, 2018. http://www.who.int/mediacentre/factsheets/fs310/en/.

116. "Why Health Care Is Ripe for Digital Disruption," *Knowledge@Wharton*, Opinion, February 3, 2017. http://knowledge.wharton.upenn.edu/article/why-health-care-is-ripe-for-digital-disruption/.

CHAPTER 7—3D PRINTING

117. "3D Printing Comes of Age in US Industrial Manufacturing," PWC website, April 2016. https://www.pwc.com/us/en/industrial-products/publications/assets/pwc-next-manufacturing-3d-printing-comes-of-age.pdf.

118. "A Printed Smile," *The Economist*, April 28, 2016. http://www.economist.com/news/science-and-technology/21697802-3d-printing-coming-age-manufacturing-technique-printed-smile.

119. "Agricultural Land (% of Land Area)," World Bank Open Data website. https://data.worldbank.org/.

120. Atkinson, Robert D., Luke A. Stewart, Scott M. Andes, and Stephen J. Ezell. "Worse Than the Great Depression: What Experts Are Missing About American Manufacturing Decline," ITIF website, March 2012. http://www2.itif.org/2012-american-manufacturing-decline.pdf.

121. Anderson, Chris. "In the Next Industrial Revolution, Atoms Are the New Bits," *Wired*, January 25, 2010. https://www.wired.com/2010/01/ff_newrevolution/.

122. Dormehl, Luke. "The Brief but Building History of 3D Printing," *Digital Trends*, February 24, 2019, Emerging Tech. https://www.digitaltrends.com/cool-tech/history-of-3d-printing-milestones/.

123. Eshel, Gidon, Alon Shepon, Tamar Makov, and Ron Milo. "Land, Irrigation Water, Greenhouse Gas, and Reactive Nitrogen Burdens of Meat, Eggs, and Dairy Production in the United States," *PNAS—Proceedings of the National Academy of Sciences of the United States of America*, July 21, 2014. http://www.pnas.org/content/111/33/11996.

124. Fallows, James. "Why the Maker Movement Matters: Part 1, the Tools Revolution," *The Atlantic*, June 5, 2016, https://www.theatlantic.com/business/archive/2016/06/why-the-maker-movement-matters-part-1-the-tools-revolution/485720/.

125. Fallows, James. "Why the Maker Movement Matters: Part 2, Agility," *The Atlantic*, June 9, 2016, Business. https://www.theatlantic.com/business/archive/2016/06/why-the-maker-movement-matters-agility/486293/.

126. "Investable Sectors: An Introduction," Food Crunch website, n.d. http://www.foodcrunch.com/precision-agriculture/.

127. Jacobs, A.J. "Dinner Is Printed," *New York Times*, September 21, 2013. http://www.nytimes.com/2013/09/22/opinion/sunday/dinner-is-printed.html.

128. Lewis, Robert. "Calling All Makers: Visit NASA Solve," NASA website, June 22, 2016. https://www.nasa.gov/feature/calling-all-makers-visit-nasa-solve.

129. Lipson, Hod, and Melba Kurman. *Fabricated: The New World of 3D Printing*. Wiley, 2013.

130. Lonjon, Capucine. "The History of 3D Printer: from Rapid Prototyping to Additive Fabrication," *Sculpteo*, March 1, 2017. https://www.sculpteo.com/blog/2017/03/01/whos-behind-the-three-main-3d-printing-technologies.

131. Lou, Nicole, and Peek, Katie. "By the Numbers: The Rise of the Makerspace," *Popular Science*, February 23, 2016, DIY. http://www.popsci.com/rise-makerspace-by-numbers.

132. Perry, Mark J. "Manufacturing's Declining Share of GDP is a Global Phenomenon, and It's Something to Celebrate," US Chamber of Commerce Foundation website, March 22, 2012. https://www.uschamberfoundation.org/blog/post/manufacturing-s-declining-share-gdp-global-phenomenon-and-it-s-something-celebrate/34261.

133. Wikipedia, s.v. "Stereolithography." https://en.wikipedia.org/wiki/Stereolithography.

134. "The Free Beginner's Guide," 3D Printing Industry website, n.d. https://3dprintingindustry.com/3d-printing-basics-free-beginners-guide/history/.

135. Thewihsen, Frank, Stefana Karevska, Alexandra Czok, Chris Pateman-Jones, and Daniel Krauss. "If 3D Printing Has Changed the Industries of Tomorrow, How Can Your Organization Get Ready Today?" Ernst & Young website, 2016. https://www.ey.com/Publication/vwLUAssets/ey-3d-printing-report/$FILE/ey-3d-printing-report.pdf.

CHAPTER 8—VIRTUAL REALITY AND VIDEO GAMES

136. Armstrong, Paul. "Just How Big Is the Virtual Reality Market and Where Is It Going Next?" *Forbes*, April 6, 2017. https://www.forbes.com/sites/paularmstrongtech/2017/04/06/just-how-big-is-the-virtual-reality-market-and-where-is-it-going-next/#176a59b74834.

137. Wikipedia, s.v. "Brookhaven National Laboratory." https://en.wikipedia.org/wiki/Brookhaven_National_Laboratory.

138. Buffum, Jude. "Bill Pitts, '68," *Stanford Magazine*, April 30, 2012. https://stanfordmag.org/contents/bill-pitts-68.

139. Coleridge, Samuel Taylor. *Biographia Literaria*. 1847. First published 1817. https://www.gutenberg.org/files/6081/6081-h/6081-h.htm.

140. Flator, Ira. *They All Laughed . . . From Light Bulbs to Lasers: The Fascinating Stories Behind the Great Inventions That Have Changed Our Lives.* Harper Collins, 1993.

141. Hamari, Juho, Jonna Koivisto, and Harri Sarsa. "Does Gamification Work?—A Literature Review of Empirical Studies on Gamification," *Research gate,* January 2014. https://www.researchgate.net/publication/256743509_Does_Gamification_Work_-_A_Literature_Review_of_Empirical_Studies_on_Gamification.

142. "History of Virtual Reality," Virtual Reality Society website, n.d. https://www.vrs.org.uk/virtual-reality/history.html.

143. Macknik, Stephen L., Mac King, James Randi, Apollo Robbins, Teller, John Thompson, and Susana Martinez-Conde. "Attention and Awareness in Stage Magic: Turning Tricks into Research," *Nature Reviews Neuroscience,* Vol. 9, pp. 871–879. July 30, 2008. https://www.nature.com/articles/nrn2473.

144. Lehrer, Jonah. "Magic and the Brain: Teller Reveals the Neuroscience of Illusion," *Wired,* April 20, 2009, Science. https://www.wired.com/2009/04/ff-neuroscienceofmagic/.

145. Lieberoth, Andreas. "Shallow Gamification: Testing Psychological Effects of Framing an Activity as a Game," *Sage Journals,* December 1, 2014, Research Article. http://journals.sagepub.com/doi/abs/10.1177/1555412014559978.

146. Lovece, Frank. "The Honest-to-Goodness History of Home Video Games." *Scribd,* June 1983. https://www.scribd.com/document/146227082/The-Honest-to-Goodness-History-of-Home-Video-Games.

147. Markoff, John. "In a Video Game, Tackling the Complexities of Protein Folding," *New York Times,* August 4, 2010 Science. http://www.nytimes.com/2010/08/05/science/05protein.html.

148. McNamara, Patrick. "Virtual Reality and Dream Research," *Psychology Today,* January 22, 2017. https://www.psychologytoday.com/blog/dream-catcher/201701/virtual-reality-and-dream-research.

149. Merel, Tim. "The Reality of VR/AR Growth," *Tech Crunch,* January 11, 2017. https://techcrunch.com/2017/01/11/the-reality-of-vrar-growth/.

150. Mann, Estle Ray. "The Baer Essentials," They Create Worlds website, n.d. https://videogamehistorian.wordpress.com/tag/estle-ray-mann/.

151. May-raz, Eran, and Daniel Lazo. *Sight* (short film). 2012. https://vimeo.com/46304267.

152. Online Etymology Dictionary. https://www.etymonline.com/.

153. Parkin, Simon. "Postscript: Ralph Baer, a Video Game Pioneer," *New Yorker,* December 8, 2014, Annals of Technology. https://www.newyorker.com/tech/annals-of-technology/postscript-ralph-baer-video-game-pioneer.

154. Pitts, Bill. "A Nutty Idea," They Create Worlds website, n.d. https://videogamehistorian.wordpress.com/tag/bill-pitts/.

155. Rovell, Darren. "427 Million People Will Be Watching Esports by 2019, Reports Newzoo," *ESPN*, May 11, 2016. http://www.espn.com/espnw/sports/article/15508214/427-million-people-watching-esports-2019-reports-newzoo.

156. Smith, Ryan P. "How the First Popular Video Game Kicked Off Generations of Virtual Adventure," *The Smithsonian Magazine*, December 13, 2018. https://www.smithsonianmag.com/smithsonian-institution/how-first-popular-video-game-kicked-off-generations-virtual-adventure-180971020/.

157. Takahashi, Dean. "Pokémon Go is the Fastest Mobile Game to Hit $600 Million in Revenues," *Venture Beat*, October 20, 2016. https://venturebeat.com/2016/10/20/pokemon-go-is-the-fastest-mobile-game-to-hit-600-million-in-revenues/.

158. "The Link Flight trainer—A Historic Mechanical Engineering Landmark," American Society of Mechanical Engineers website, New York, June 10, 2000. https://www.asme.org/wwwasmeorg/media/ResourceFiles/AboutASME/Who%20We%20Are/Engineering%20History/Landmarks/210-Link-C-3-Flight-Trainer.pdf.

159. "The Theme Park of the Future Could Be in This Chinese Basement," *Bloomberg*, April 3, 2017, Technology. https://www.bloomberg.com/news/articles/2017-04-03/this-could-be-the-most-fun-you-ll-have-in-an-empty-basement.

160. Wikipedia, s.v. "Virtual reality sickness." https://en.wikipedia.org/wiki/Virtual_reality_sickness.

161. Weinbaum, Stanley. *Pygmalion's Spectacles*. 1935. https://www.gutenberg.org/files/22893/22893-h/22893-h.htm.

162. Wikipedia, s.v. "William Higinbotham." https://en.wikipedia.org/wiki/William_Higinbotham.

163. Wong, Kevin. "The Forgotten History of 'The Oregon Trail,' as Told by Its Creators," *Motherboard Tech by Vice*, February 15, 2017. https://motherboard.vice.com/en_us/article/qkx8vw/the-forgotten-history-of-the-oregon-trail-as-told-by-its-creators.

CHAPTER 9—EDUCATION

164. Calderon, Angel J. "Massification of higher education revisited," *Academia*, RMT University, June 2018. https://www.academia.edu/36975860/Massification_of_higher_education_revisited.

165. Carr, Nicholas. *The Shallows: What the Internet Is Doing to Our Brains*. W. W. Norton & Company, 2011.

166. "Global Education Technology Market to Reach $341B by 2025," *Markets Insider* Press Release, PR Newswire, January 24, 2019. https://markets.businessinsider.com/news/stocks/global-education-technology-market-to-reach-341b-by-2025-1027892295.

167. Henkel, Linda A. "Point-and-Shoot Memories: The Influence of Taking Photos on Memory for a Museum Tour," *Sage Journals*, December 5, 2013. Psychological Science. http://journals.sagepub.com/doi/abs/10.1177/0956797613504438.

168. HolonIQ. "$87bn of Global EdTech Funding predicted through 2030; $43bn last decade," Cision PR Newswire, January 28, 2020. https://www.prnewswire.com/news-releases/87bn-of-global-edtech-funding-predicted-through-2030-32bn-last-decade-300994266.html.

169. "Memory Loss Causes: Taking Pictures May Ruin What You Recall," HuffPost, December 10, 2013, Living. http://www.huffingtonpost.ca/2013/12/10/memory-loss-causes_n_4419560.html.

170. Nicholas Carr Interview. "'The Shallows': This Is Your Brain Online," NPR, June 2, 2010. Author Interviews. http://www.npr.org/templates/story/story.php?storyId=127370598.

171. Roser, Max, and Esteban Ortiz-Ospina. "Tertiary Education," Our World in Data website, n.d. https://ourworldindata.org/tertiary-education.

172. Small, G.W., T.D. Moody, P. Siddarth, and S.Y. Bookheimer. "Your Brain on Google: Patterns of Cerebral Activation During Internet Searching," *American Journal of Geriatric Psychiatry*, Vol. 17, Issue 2, February 2009, Pages 116–126. https://www.ncbi.nlm.nih.gov/pubmed/19155745.

173. UNESCO. "Education: Expenditure on Education as % of GDP (from Government Sources)," UNESCO Institute for Statistics website, n.d. http://data.uis.unesco.org/?queryid=181.

174. UNESCO. "Six Ways to Ensure Higher Education Leaves No One Behind," UNESCO documents website, 2017. http://unesdoc.unesco.org/images/0024/002478/247862E.pdf.

CHAPTER 10—SOCIAL NETWORKS

175. Barakat, Zena, and Sean Patrick Farrell. "The Dawn of Computer Love," *New York Times*, February 2013, Technology. https://www.nytimes.com/video/technology/100000002063332/the-dawn-of-computer-love.html.

176. Wikipedia, s.v. "Book Stacks Unlimited." https://en.wikipedia.org/wiki/Book_Stacks_Unlimited.

177. Ferster, C.B., and B.F. Skinner. *Schedules of Reinforcement*. Copley Publishing Group, 1997. First published 1957.

178. Ensmenger, Nathan. "Computer Dating in the 1960s," The Computer Boys Take Over website, March 5, 2014. http://thecomputerboys.com/?p=654.

179. "Global Digital Report 2019," We Are Social website. https://digitalreport. wearesocial.com/.

180. Hvistendahl, Mara. "Inside China's Vast New Experiment in Social Ranking," *Wired*, December 14, 2017. https://www.wired.com/story/age-of-social-credit/.

181. "Internet Growth Statistics," Internet World Stats website, n.d. https://www. internetworldstats.com/emarketing.htm.

182. Klein, Dustin S. "Visionary in Obscurity: Charles Stack," *Smart Business*, July 22, 2002, Cleveland. http://www.sbnonline.com/article/visionary-in-obscurity-charles-stack-operates-in-two-business-communities-151-cleveland-and-the-internet-151-and-isn-146-t-well-known-in-either-this-time-around-that-146-s-going-to-change-he-hopes/.

183. Lipsman, Andrew. "Global Ecommerce 2019: Ecommerce Continues Strong Gains Amid Global Economic Uncertainty," *eMarketer*, June 27, 2019. https:// www.emarketer.com/content/global-ecommerce-2019.

184. Markoff, John. *What the Dormouse Said: How the Sixties Counterculture Shaped the Personal Computer Industry*. Penguin Books, 2006.

185. Mc Mahon, Ciarán. "Why Do We 'Like' Social Media?" *The Psychologist*, September 2015. https://thepsychologist.bps.org.uk/volume-28/september-2015/why-do-we-social-media.

186. "Mental Health," World Health Organization website, n.d. https://www.who. int/mental_health/prevention/suicide/suicideprevent/en/.

187. Metcalfe, Robert. "Guest Blogger Bob Metcalfe: Metcalfe's Law Recurses Down the Long Tail of Social Networks," VCMike's blog, August 18, 2006. https://vcmike.wordpress.com/2006/08/18/metcalfe-social-networks/.

188. Wikipedia, s.v. "Michael Aldrich." https://en.wikipedia.org/wiki/Michael_Aldrich.

189. "money (n.)," entry in the Online Etymology Dictionary, 2020. https://www. etymonline.com/word/money.

190. Ortega, Josué, and Philipp Hergovich. "The Strength of Absent Ties: Social Integration via Online Dating." *arXiv*, September 29, 2017. https://arxiv.org/abs/1709.10478.

191. Pennisi, Elizabeth. "How Humans Became Social," *Wired*, November 9, 2011, Science Now. https://www.wired.com/2011/11/humans-social/.

192. Philip Fialer. "A Life Well-Lived," *Legacy.com—San Francisco Chronicle*, January 5, 2014. https://www.legacy.com/obituaries/sfgate/obituary.aspx?n=philip-fialer&pid=168884439.

193. Sherman, Lauren E., Ashley A. Payton, Leanna M. Hernandez, Patricia M. Greenfield, and Mirella Dapretto. "The Power of the *Like* in Adolescence: Effects of Peer Influence on Neural and Behavioral Responses to Social Media," *Psychological Science*, May 31, 2016. https://journals.sagepub.com/doi/abs/10.1177/0956797616645673.

194. Shultz, Susanne, Christopher Opie, and Quentin D. Atkinson. "Stepwise Evolution of Stable Sociality in Primates," *Nature*, Vol. 479, pp. 219–222. November 9, 2011. https://www.nature.com/articles/nature10601.

195. Slater, Dan. "A Million First Dates," *The Atlantic*, February 2013. https://www.theatlantic.com/magazine/archive/2013/01/a-million-first-dates/309195/.

196. "Suicide Prevention," Centers for Disease Control and Prevention website, n.d. https://www.cdc.gov/violenceprevention/suicide/index.html.

197. Norman, Jeremy. "The First Computer Matching Dating Service," History of Information website, n.d. http://www.historyofinformation.com/detail.php?entryid=3970.

198. Winterman, Denise, and Jon Kelly. "Online Shopping: The Pensioner Who Pioneered a Home Shopping Revolution," *BBC News Magazine*, September 16, 2013. http://www.bbc.com/news/magazine-24091393.

199. Zhang, Xing-Zhou, Jing-Jie Liu, and Zhi-Wei Xu. "Tencent and Facebook Data Validate Metcalfe's Law," *Springer Link— Journal of Computer Science and Technology,* Volume 30, Issue 2, pp 246–251. March 13, 2015. https://link.springer.com/article/10.1007%2Fs11390-015-1518-1.

CHAPTER 11—FINTECH AND CRYPTOCURRENCIES

200. Ferguson, Niall. *The Ascent of Money: A Financial History of the World*. Penguin Group, 2008.

201. "First Paper Money," Guinness World Records website, n.d. http://www.guinnessworldrecords.com/world-records/first-paper-money.

202. "The Global Findex Database 2017," The World Bank website, n.d. https://globalfindex.worldbank.org/.

203. Goldsborough, Reid. "A Case for the World's Oldest Coin: Lydian Lion," 2013. http://rg.ancients.info/lion/article.html.

204. Wikipedia, s.v. "History of money." https://en.wikipedia.org/wiki/History_of_money.

205. Wikipedia, s.v. "Lydians." https://en.wikipedia.org/wiki/Lydians.

206. McKinsey & Company. "Global payments Report 2019: Amid sustained growth, accelerating challenges demand bold actions," McKinsey & Co Global Banking Practice, September 2019. https://www.mckinsey.com/~/media/ mckinsey/industries/financial%20services/our%20insights/tracking%20the%20 sources%20of%20robust%20payments%20growth%20mckinsey%20global%20 payments%20map/global-payments-report-2019-amid-sustained-growth-vf. ashx.

207. Pollari, Ian, and Anton Ruddenklau. "Pulse of Fintech H1 2020," KPMG, September 2020. https://assets.kpmg/content/dam/kpmg/xx/pdf/2020/09/ pulse-of-fintech-h1-2020.pdf.

208. "World Payments Report 2020," Capgemini, October 6, 2020. https:// worldpaymentsreport.com/.

CHAPTER 12—ENCRYPTION AND BLOCKCHAIN

209. "A Penny Here, a Penny There," *The Economist*, May 7, 2015, Special report, https://www.economist.com/news/special-report/21650297-if-you-have-moneyand-even-if-you-dontyou-can-now-pay-your-purchases-myriad-ways.

210. "All Cryptocurrencies," CoinMarketCap website, n.d. https://coinmarketcap. com/all/views/all/.

211. "Blockchain Futures Lab," Institute for the Future website, n.d. http://www.iftf. org/blockchainfutureslab.

212. Browne, Ryan. "Digital Payments Expected to Hit 726 Billion by 2020—But Cash Isn't Going Anywhere Yet," *CNBC*, October 9, 2017. https://www.cnbc. com/2017/10/09/digital-payments-expected-to-hit-726-billion-by-2020-study-finds.html.

213. Catalini, Christian, and Joshua S. Gans. "Some Simple Economics of the Blockchain," *SSRN*, November 27, 2016. https://papers.ssrn.com/sol3/papers. cfm?abstract_id=2874598.

214. Wikipedia, s.v. "Clifford Cocks." https://en.wikipedia.org/wiki/Clifford_Cocks.

215. El-Abbadi, Mostafa, "The Fate of the Library of Alexandria," Encyclopædia Britannica. https://www.britannica.com/topic/Library-of-Alexandria/The-fate-of-the-Library-of-Alexandria.

216. Hackett, Robert. "Walmart and 9 Food Giants Team Up on IBM Blockchain Plans," *Fortune,* August 22, 2017. http://fortune.com/2017/08/22/walmart-blockchain-ibm-food-nestle-unilever-tyson-dole/.

217. Wikipedia, s.v. "History of cryptography." https://en.wikipedia.org/wiki/ History_of_cryptography.

218. "Hyperledger Members," n.d. https://www.hyperledger.org/members.

219. Wikipedia, s.v. "Key exchange." https://en.wikipedia.org/wiki/Key_exchange.

220. Malmo, Christopher. "A Single Bitcoin Transaction Takes Thousands of Times More Energy Than a Credit Card Swipe," *Vice,* March 7, 2017. Motherboard Tech by Vice. https://motherboard.vice.com/en_us/article/ypkp3y/bitcoin-is-still-unsustainable.

221. Orcutt, Mike. "How Secure Is Blockchain Really?" *MIT Technology Review,* April 25, 2018. https://www.technologyreview.com/s/610836/how-secure-is-blockchain-really/.

222. Orcutt, Mike. "Once Hailed as Unhackable, Blockchain Are Now Getting Hacked," *MIT Technology Review,* February 19, 2019. https://www.technologyreview.com/s/612974/once-hailed-as-unhackable-blockchains-are-now-getting-hacked/.

223. Rizzo, Pete. "World Economic Forum Survey Projects Blockchain 'Tipping Point' by 2023," *Coindesk,* September 24, 2015. https://www.coindesk.com/world-economic-forum-governments-blockchain.

224. Roberts, Jeff John. "The Diamond Industry Is Obsessed with the Blockchain," *Fortune,* September 12, 2017. http://fortune.com/2017/09/12/diamond-blockchain-everledger/.

225. Schatsky, David, and Linda Pawczuk. "Deloitte Survey: Blockchain Reaches Beyond Financial Services with Some Industries Moving Faster," Deloitte website, New York, December 13, 2016. https://www2.deloitte.com/us/en/pages/about-deloitte/articles/press-releases/deloitte-survey-blockchain-reaches-beyond-financial-services-with-some-industries-moving-faster.html.

226. Wikipedia, s.v. "RSA (cryptosystem)." https://en.wikipedia.org/wiki/RSA_(cryptosystem).

227. World Trade Organization. "Highlights of world trade in 2019," in *World Trade Statistical Review 2020,* World Trade Organization. https://www.wto.org/english/res_e/statis_e/wts2020_e/wts2020chapter02_e.pdf.

CHAPTER 13—ROBOTICS

228. Bachman, Justin. "The Lonely Future of Buying Stuff," *Bloomberg,* September 13, 2017. https://www.bloomberg.com/features/2017-future-of-automation/.

229. Wikipedia, s.v. "Ctesibius." https://en.wikipedia.org/wiki/Ctesibius.

230. Desjardins, Jeff. "The Emergence of Commercial Drones," *Visual Capitalist,* December 14, 2016. http://www.visualcapitalist.com/emergence-commercial-drones/.

231. "The Drone Market Report 2020–2025," Drone Industry Insights, 2020. https://www.droneii.com/project/drone-market-report-2020-2025.

232. FAA. "UAS by the Numbers," Federal Aviation Administration website, September 2020. https://www.faa.gov/uas/resources/by_the_numbers/.

233. "Friedrich Kaufmann," History-Computer.com, n.d. https://history-computer. com/Dreamers/Kaufmann.html.

234. Furukawa, Keiichi. "Honda's Asimo robot bows out but finds new life," *Nikkei Asia*, June 28, 2018. https://asia.nikkei.com/Business/Companies/Honda-s-Asimo-robot-bows-out-but-finds-new-life.

235. "Computers Gone Wild: Impact and Implications of Developments in Artificial Intelligence on Society," Guest blogger on Future of Life Institute website, May 6, 2016. https://futureoflife.org/2016/05/06/computers-gone-wild/.

236. Hardesty, Larry. "Drones relay RFID signals for inventory control," *MIT News*, August 25, 2017. https://news.mit.edu/2017/drones-relay-rfid-signals-inventory-control-0825.

237. "History of the Robotics Institute," Carnegie Mellon University website, n.d. https://www.ri.cmu.edu/about/ri-history/.

238. History.com Editors. "Joseph Kennedy JR," website for history.com, August 21, 2018. https://www.history.com/topics/1960s/joseph-kennedy-jr.

239. Hoggett, Reuben. "1928—'Gatukentosku' Pneumatic Writing Robot— Makoto Nishimura (Japanese)," *CyberneticZoo,* n.d. http://cyberneticzoo. com/robots/1928-gakutensoku-pneumatic-writing-robot-makoto-nishimura-japanese/.

240. Hoggett, Reuben. "1937—Elektro—Joseph M. Barnett (American)," *CyberneticZoo*, n.d. http://cyberneticzoo.com/robots/1937-elektro-joseph-m-barnett-american/.

241. IFR. "Welcome to the IFR Press Conference, 18th September 2019, Shanghai," International Federal of Robotics, September 2019. https://ifr.org/downloads/press2018/IFR%20World%20Robotics%20Presentation%20-%2018%20Sept%202019.pdf.

242. Wikipedia, s.v. "Ismail al-Jazari." https://en.wikipedia.org/wiki/Ismail_al-Jazari.

243. Kubota, Taylor. "Stanford's robotics legacy," *Stanford News*, January 16, 2019. https://news.stanford.edu/2019/01/16/stanfords-robotics-legacy/.

244. Lumb, David. "MIT Researchers Use Drone Fleets to Track Warehouse Inventory," *Engadget,* August 25, 2017. https://www.engadget. com/2017/08/25/mit-drone-fleets-track-warehouse-inventory/.

245. Wikipedia, s.v. "Norbert Wiener." https://en.wikipedia.org/wiki/Norbert_Wiener.

246. O'Brien, Matt. "As robots take over warehousing, workers pushed to adapt," *AP News*, December 30, 2019. https://apnews.com/056b44f5bfff11208847aa9768f10757.

247. Parsons, Mark. "Automation and Robotics: The Supply Chain of the Future," *Supply Chain Digital,* May 3, 2017. http://www.supplychaindigital.com/technology/automation-and-robotics-supply-chain-future.

248. "Richards Family Robot Archive. Who Was He?" 2018. http://www.richardsrobots.com/captain-wh-richards.html.

249. Tesla, Nikola. *My Inventions: The Autobiography of Nikola Tesla.* Hart Brothers Pub, 1982. First published 1921.

250. "The Curious Origin of the Word 'Robot'," Interesting Literature website, March 14, 2016. https://interestingliterature.com/2016/03/14/the-curious-origin-of-the-word-robot/.

251. "The Dronefather," *The Economist,* December 1, 2012. https://www.economist.com/news/technology-quarterly/21567205-abe-karem-created-robotic-plane-transformed-way-modern-warfare.

252. Wikipedia, s.v. "The Turk." https://en.wikipedia.org/wiki/The_Turk.

253. Turi, Jon. "Tesla's Boat: A Drone Before Its Time," *Engadget,* January 19, 2014. https://www.engadget.com/2014/01/19/nikola-teslas-remote-control-boat/.

254. Vyas, Kashyap. "A Brief History of Drones: The Remote Controller Unmanned Aerial Vehicles (UAVs)," *Interesting Engineering,* January 2, 2018. https://interestingengineering.com/a-brief-history-of-drones-the-remote-controlled-unmanned-aerial-vehicles-uavs.

255. Wiener, Norbert. *The Human Use of Human Beings.* Da Capo Press, 1988. First published 1950.

256. Wilson, Edward. "Thank You Vasili Arkhipov, the Man Who Stopped Nuclear War," *The Guardian,* October 27, 2012, Opinion. https://www.theguardian.com/commentisfree/2012/oct/27/vasili-arkhipov-stopped-nuclear-war.

257. "World Trade Statistical Review 2016," World Trade Organization website, 2016. https://www.wto.org/english/res_e/statis_e/wts2016_e/wts2016_e.pdf.

258. "BrandZ Global Top 100 Infographic," brandz.com website, June 30, 2020. https://www.brandz.com/admin/uploads/files/2020_BrandZ_Global_Top_100_Infographic.pdf.

CHAPTER 14—NANOTECHNOLOGY

259. Feynman, Richard P. "There's Plenty of Room at the Bottom," Caltech Library website, n.d. http://calteches.library.caltech.edu/47/2/1960Bottom.pdf.

260. Kornei, Katherine. "The Beginning of Nanotechnology at the 1959 APS Meeting," *APS News,* November 2016. https://www.aps.org/publications/apsnews/201611/nanotechnology.cfm.

261. Laboratório Nacional de Nanotecnologia Aplicada ao Agronegócio (LNNA). LNNA. n.p., n.d. https://www.agropediabrasilis.cnptia.embrapa.br/web/agronano-rede/lnna.

262. Lavan, D.A., T. McGuire, and R. Langer. "Small-Scale Systems for In Vivo Drug Delivery," US National Library of Medicine website, October 21, 2003. https://www.ncbi.nlm.nih.gov/pubmed?uid=14520404&cmd=showdetailview.

263. Lehr, Dick. "The Racist Legacy of Woodrow Wilson," *The Atlantic*, November 27, 2015. https://www.theatlantic.com/politics/archive/2015/11/wilson-legacy-racism/417549/.

264. Wikipedia, s.v. "Nanomedicine." https://en.wikipedia.org/wiki/Nanomedicine.

265. "NNI Budget," National Nanotechnology Initiative (NNI) website, n.d. https://www.nano.gov/about-nni/what/funding.

266. Nanotechnology Products Database. ca. 2019. https://product.statnano.com/.

267. Wikipedia, s.v. "Norio Taniguchi." https://en.wikipedia.org/wiki/Norio_Taniguchi.

268. Office of Communication. "President Eisgruber's message to community on removal of Woodrow Wilson name from public policy school and Wilson College," Princeton University, June 27, 2020. https://www.princeton.edu/news/2020/06/27/president-eisgrubers-message-community-removal-woodrow-wilson-name-public-policy.

269. StatNano. ca. 2019. http://statnano.com/.

270. "UNESCO Science Report Towards 2030. Nanotechnology Is a Growing Research Priority," UNESCO website, July 12, 2016. Natural Sciences Sector. http://www.unesco.org/new/en/natural-sciences/science-technology/single-view-sc-policy/news/nanotechnology_is_a_growing_research_priority/.

271. Vance, Marina E., Todd Kuiken, Eric P. Vejerano, Sean P. McGinnis, Michael F. Hochella Jr., David Rejeski, and Michael S. Hull. "Nanotechnology in the Real World: Redeveloping the Nanomaterial Consumer Products Inventory," *Beilstein Journal of Nanotechnology*, August 21, 2015. https://www.beilstein-journals.org/bjnano/articles/6/181.

CHAPTER 15—AIRPLANES, ROCKETS, AND SATELLITES

272. Wikipedia, s.v. "Balance of terror." https://en.wikipedia.org/wiki/Balance_of_terror.

273. "Center for Near-Earth Object Studies," NASA website, ca. 2019. https://cneos.jpl.nasa.gov/.

274. Davenport, Christian. "The Inside Story of How Billionaires Are Racing to Take You to Outer Space," *Washington Post*, August 19, 2016, Business. https://www.washingtonpost.com/business/economy/the-billionaire-space-barons-and-the-next-giant-leap/2016/08/19/795a4012-6307-11e6-8b27-bb8ba39497a2_story.html?noredirect=on&utm_term=.6e078b20f006.

275. Wikipedia, s.v. "Drake equation." https://en.wikipedia.org/wiki/Drake_equation.

276. Godwin, Matthew. "The Cold War and the Early Space Race," *History in Focus*, n.d. https://www.history.ac.uk/ihr/Focus/cold/articles/godwin.html.

277. Grady, Monica. "A Handshake in Space Changed US-Russia Relations: How Long Will It Last?" *The Conversation*, July 17, 2015. https://theconversation.com/a-handshake-in-space-changed-us-russia-relations-how-long-will-it-last-44846.

278. Grush, Loren. "Why Defining the Boundary of Space May Be Crucial for the Future of Spaceflight," *The Verge*, December 13, 2018, Science. https://www.theverge.com/2018/12/13/18130973/space-karman-line-definition-boundary-atmosphere-astronauts.

279. Kovitch. "Fly to New York in 20 Minutes: Transatlantic Travel 1920–2020," DialaFlight website, September 28, 2012. https://www.dialaflight.com/flights-to-new-york/.

280. Lavender, Andrew. "How many satellites are orbiting the Earth in 2020?" Pixalytics, May 27, 2020. https://www.pixalytics.com/satellites-orbiting-earth-2020/.

281. "OSCAR I and Amateur Radio Satellites: Celebrating 50 Years," The National Association for Amateur Radio (ARRL) website, November 5, 2011. http://www.arrl.org/news/oscar-i-and-amateur-radio-satellites-celebrating-50-years.

282. Satellite Industry Association. "2020 State of the Satellite Industry, Report—Two Page Summary," 2020. https://sia.org/news-resources/state-of-the-satellite-industry-report/.

283. Wikipedia, s.v. "Timeline of private spaceflight." https://en.wikipedia.org/wiki/Timeline_of_private_spaceflight.

284. Thomas, Zoe. "The woman who paid $250,000 to go into space," *BBC News*, January 12, 2020. https://www.bbc.com/news/business-50929064.

285. Union of Concerned Scientists. ca. 2019. https://www.ucsusa.org/.

286. "United Nations Register of Objects Launched into Outer Space," United Nations Office for Outer Space Affairs website, ca. 2019. http://www.unoosa.org/oosa/en/spaceobjectregister/index.html.

287. Verne, Jules. *The Begum's Fortune*. Wildside Press, 2003. First published 1879.

288. XPRIZE. ca. 2019. https://www.xprize.org/.

289. Jones, Harry W., The Recent Large Reduction in Space Launch Cost, 48th International Conference on Environmental Systems, 2018.

290. https://ttu-ir.tdl.org/bitstream/handle/2346/74082/ICES_2018_81. pdf?sequence=1&isAllowed=y

CHAPTER 16—ENERGY

291. Andrae, Anders S.G. and Tomas Edler. "On Global Electricity Usage of Communication Technology: Trends to 2030," *MDPI*, February 27, 2015. https://www.mdpi.com/2078-1547/6/1/117/htm.

292. Arman, Shehabi, Sarah Josephine Smith, Dale A. Sartor, Richard E. Brown, Magnus Herrlin, Jonathan G. Koomey, Eric R. Masanet, Nathaniel Horner, Inês Lima Azevedo, and William Lintner. "United States Data Center Energy Usage Report," Berkeley lab, Energy Technologies Area website, June 2016. https://eta.lbl.gov/publications/united-states-data-center-energy.

293. Chandler, David L. "Vast Amounts of Solar Energy Radiate to the Earth, But Tapping It Cost-Effectively Remains a Challenge," *Phys.org*, October 26, 2011. https://phys.org/news/2011-10-vast-amounts-solar-energy-earth.html.

294. Caldeira, Ken, Atul K. Jain, and Martin I. Hoffert. "Climate Sensitivity Uncertainty and the Need for Energy Without CO_2 Emission," *Science*, Vol. 299, Issue 5615, pp. 2052–2054. March 28, 2003. https://science.sciencemag. org/content/299/5615/2052.

295. "Clicking Clean Report 2017," Greenpeace website. http://www.greenpeace. org/usa/global-warming/click-clean.

296. "Climate Change Performance Index," GermanWatch website, n.d. https:// www.climate-change-performance-index.org/.

297. "Digitalization & Energy," International Energy Agency (IEA). IEA website. n.d. https://www.iea.org/publications/freepublications/publication/ DigitalizationandEnergy3.pdf.

298. "Energy Statistics Pocketbook 2018," Department of Economic and Social Affairs. United Nations website, 2018. https://unstats.un.org/unsd/energy/ pocket/2018/2018pb-web.pdf.

299. "Fossil CO2 and GHG emissions of all world countries, 2019 report," Emissions Database for Global Atmospheric Research (EDGAR), European Commission. https://edgar.jrc.ec.europa.eu/overview.php?v=booklet2019.

300. Gamble, Chris, and Jim Gao. "Safety-First AI for Autonomous Data Centre Cooling and Industrial Control," *DeepMind*, August 17, 2018. https:// deepmind.com/blog/safety-first-ai-autonomous-data-centre-cooling-and-industrial-control/.

301. "Global Energy Statistical Yearbook 2019," Enerdata website, ca. 2019. https:// yearbook.enerdata.net/.

302. Wikipedia, s.v. "Hornsdale Power Reserve." https://en.wikipedia.org/wiki/Hornsdale_Power_Reserve.

303. IEA. "Global energy investment in 2018," IEA website, December 3, 2019. https://www.iea.org/data-and-statistics/charts/global-energy-investment-in-2018.

304. Jones, Nicola. "How to Stop Data Centres from Gobbling Up the World's Electricity," *Nature*, September 12, 2018, News Feature. https://www.nature.com/articles/d41586-018-06610-y.

305. Leuven, KU. "Generating Power from Polluted Air," *Phys.org*, May 8, 2017. https://phys.org/news/2017-05-power-polluted-air.html.

306. Lindeman, Todd. "1.3 Billion Are Living in the Dark," *Washington Post*, November 6, 2015. https://www.washingtonpost.com/graphics/world/world-without-power/?noredirect=on.

307. Wikipedia, s.v. "List of countries by electricity consumption." https://en.wikipedia.org/wiki/List_of_countries_by_electricity_consumption.

308. Masanet, Eric, and Nuoa Lei. "How Much Energy Do Data Centers Really Use?" Energy Innovation Policy & Technology website, March 17, 2020. https://energyinnovation.org/2020/03/17/how-much-energy-do-data-centers-really-use/.

309. Masanet, Eric, Arman Shehabi, Nuoa Lei, Sarah Smith, and Jonathan Koomey. "Recalibrating global data center energy-use estimates," *Science*, February 28, 2020. https://science.sciencemag.org/content/367/6481/984/tab-article-info.

310. Rincon, Paul. "Nobel Chemistry Prize: Lithium-Ion Battery Scientists Honoured," *BBC News*, October 9, 2019. https://www.bbc.com/news/science-environment-49962133.

311. Ross, Andrew. "A Perfect Storm: The Environmental Impact of Data Centres," *Information Age*, September 17, 2018. https://www.information-age.com/a-perfect-storm-the-environmental-impact-of-data-centres-123474834/.

312. Salisbury, David. "Device Could Use Your Motion to Charge Phone," *Futurity*, July 24, 2017. http://www.futurity.org/device-energy-human-motion-1492932/.

313. Shehabi, Arman, Sarah Smith, Dale Sartor, Richard Brown, Magnus Herrlin, Jonathan Koomey, Eric Masanet, Nathaniel Horner, Inês Azevedo, and William Lintner. "United States Data Center Energy Usage Report," Lawrence Berkeley National Laboratory website. June 2016. http://eta-publications.lbl.gov/sites/default/files/lbnl-1005775_v2.pdf.

314. "Sources of Greenhouse Gas Emissions," United States Environmental Protection Agency (EPA) website, n.d. https://www.epa.gov/ghgemissions/sources-greenhouse-gas-emissions.

315. Wikipedia, s.v. "Special Report on Global Warming of 1.5°C." https://en.wikipedia.org/wiki/Special_Report_on_Global_Warming_of_1.5_%C2%B0C.

316. "Statistical Review of World Energy: 2020, 69th edition," BP website, June 2020. https://www.bp.com/content/dam/bp/business-sites/en/global/corporate/pdfs/energy-economics/statistical-review/bp-stats-review-2020-full-report.pdf.

317. Temple, James. "At This Rate, It's Going to Take Nearly 400 Years to Transform the Energy System," *MIT Technology Review,* March 14, 2018. https://www.technologyreview.com/s/610457/at-this-rate-its-going-to-take-nearly-400-years-to-transform-the-energy-system/.

318. The Intergovernmental Panel on Climate Change. IPCC. ca. 2019. https://www.ipcc.ch/.

319. Titcomb, James. "Elon Musk Says He Will Fix South Australia's Power in 100 Days—Or Do It for Free," *Telegraph,* March 10, 2017, Technology Science. http://www.telegraph.co.uk/technology/2017/03/10/elon-musk-makes-audacious-twitter-bet-fix-south-australias-power/.

320. Tollefson, Jeff. "IPCC Says Limiting Global Warming to 1.5°C Will Require Drastic Action," *Nature,* October 8, 2018. https://www.nature.com/articles/d41586-018-06876-2.

321. "Total Surface Area Required to Fuel the World with Solar," Land Art Generator website. August 13, 2009. https://landartgenerator.org/blagi/archives/127.

322. Wang, T. "Global Electricity Consumption 1980–2016," *Statista,* June 27, 2016. https://www.statista.com/statistics/280704/world-power-consumption/.

323. Weiser, Matt. "The Hydropower Paradox: Is This Energy as Clean as It Seems?" *The Guardian,* November 6, 2016. https://www.theguardian.com/sustainable-business/2016/nov/06/hydropower-hydroelectricity-methane-clean-climate-change-study.

324. Wikipedia, s.v. "World energy consumption." https://en.wikipedia.org/wiki/World_energy_consumption.

325. "World Energy Investment 2017," IEA website, n.d. https://www.iea.org/publications/wei2017/.

326. "World Energy Model," Shell website, n.d. https://www.shell.com/energy-and-innovation/the-energy-future/scenarios/shell-scenarios-energy-models/world-energy-model.html.

327. "World Energy Outlook 2018. Introduction," IEA website, ca. 2018. https://www.iea.org/weo2018/electricity/.

CHAPTER 17—NEW MATERIALS

328. Wikipedia, s.v. "Aristarchus of Samos." https://en.wikipedia.org/wiki/Aristarchus_of_Samos.

329. "Atanasoff-Berry Computer," Iowa State University website, ca. 2011. http://jva.cs.iastate.edu/operation.php.

330. Bates, Mary. "How Does a Battery Work?" MIT, School of Engineering website, May 1, 2012. https://engineering.mit.edu/engage/ask-an-engineer/how-does-a-battery-work/.

331. Bernstein, Peter. "The Top Bell Labs Innovations—Part I: The Game-Changers," *TMCnet*, August 29, 2011. http://blog.tmcnet.com/next-generation-communications/2011/08/the-top-bell-labs-innovations---part-i-the-game-changers.html.

332. Wikipedia, s.v. "Bi Sheng." https://en.wikipedia.org/wiki/Bi_Sheng.

333. "Birth of Electrochemistry," The Electrochemical Society website, n.d. https://www.electrochem.org/birth-of-electrochemistry.

334. "Colossus Computer of Max Newman and Tommy Flowers," History-Computer.com website, n.d. https://history-computer.com/ModernComputer/Electronic/Colossus.html.

335. Wikipedia, s.v. "Computer (job description)." https://en.wikipedia.org/wiki/Computer_(job_description).

336. "ENIAC," History-Computer.com website, n.d. https://history-computer.com/ModernComputer/Electronic/ENIAC.html.

337. "François Fresneau," Bouncing Balls website, n.d. http://www.bouncing-balls.com/timeline/people/nr_fresneau.htm.

338. Harris, James, and David Webb. "Achieving Optimum LED Performance with Quantum Dots," *Photonics Media*, n.d. https://www.photonics.com/a60882/Achieving_Optimum_LED_Performance_With_Quantum.

339. Wikipedia, s.v. "Johannes Gutenberg." https://en.wikipedia.org/wiki/Johannes_Gutenberg.

340. Maxwell, James. *Experiments on Colour.* Jim Worthey Lighting and Color Research website, 1855. http://www.jimworthey.com/archive/Maxwell_1855_OCRtext.pdf.

341. "Oldest Tool Use and Meat-Eating Revealed," National History Museum (Web Archive). ca. 2010. https://web.archive.org/web/20100818123718/http://www.nhm.ac.uk/about-us/news/2010/august/oldest-tool-use-and-meat-eating-revealed75831.html.

342. Wikipedia, s.v. "Point-contact transistor." https://en.wikipedia.org/wiki/Point-contact_transistor.

343. Wikipedia, s.v. "Polytetrafluoroethylene." https://en.wikipedia.org/wiki/Polytetrafluoroethylene.

344. Quinlan, Heather. "Aerogel History," *How Stuff Works*, n.d., Science. https://science.howstuffworks.com/aerogel1.htm.

345. Somma, Ann Marie. "Charles Goodyear and the Vulcanization of Rubber," *Connecticut History*, December 29, 2014. https://connecticuthistory.org/charles-goodyear-and-the-vulcanization-of-rubber/.

346. "The ABC of John Atanasoff and Clifford Berry," History-Computer.com website, n.d. https://history-computer.com/ModernComputer/Electronic/Atanasoff.html.

347. "The Element Germanium," Jefferson Lab website, n.d. https://education.jlab.org/itselemental/ele032.html.

348. "The Element Selenium," Jefferson Lab website, n.d. https://education.jlab.org/itselemental/ele034.html.

349. "The Man Who 'changed the world forever'," King's College London website, June 1, 2017. https://www.kcl.ac.uk/news/spotlight-article?id=360468c1-1909-45ea-9a2d-4c7e1e9c809b.

350. Wikipedia, s.v. "Transistor." https://en.wikipedia.org/wiki/Transistor.

351. Yardley, William. "George H. Heilmeier, and Inventor of LCDs, Dies at 77," *New York Times*, May 6, 2014. https://www.nytimes.com/2014/05/06/technology/george-h-heilmeier-an-inventor-of-lcds-dies-at-77.html.

CHAPTER 18—BIG DATA

352. "Cisco Visual Networking Index: Forecast and Trends, 2017–2022," CISCO website, ca. 2017. https://www.cisco.com/c/en/us/solutions/collateral/service-provider/visual-networking-index-vni/white-paper-c11-741490.pdf.

353. "Cisco Visual Networking Index: Global Mobile Data Traffic Forecast Update, 2017–2022 White Paper," CISCO website, February 18, 2019. https://www.cisco.com/c/en/us/solutions/collateral/service-provider/visual-networking-index-vni/mobile-white-paper-c11-520862.html.

354. Desjardins, Jeff. "This Is What Happens in a Minute on the Internet," *World Economic Forum*, March 15, 2019. https://www.weforum.org/agenda/2019/03/what-happens-in-an-internet-minute-in-2019/.

355. Duhigg, Charles. *The Power of Habit*. Random House, 2012.

356. Eisenstein, Elizabeth L. *The Printing Press as an Agent of Change*. Cambridge University Press, 1979.

357. Wikipedia, s.v. "General Conference on Weights and Measures." https://en.wikipedia.org/wiki/General_Conference_on_Weights_and_Measures.

358. Wikipedia, s.v. "General Data Protection Regulation." https://en.wikipedia.org/wiki/General_Data_Protection_Regulation.

359. Ginsberg, Jeremy, Matthew H. Mohebbi, Rajan S. Patel, Lynnette Brammer, Mark S. Smolinski, and Larry Brilliant. "Detecting Influenza Epidemics Using Search Engine Query Data," *Nature,* Vol. 457, pp. 1012–1014. February 1, 2019. https://www.nature.com/articles/nature07634.

360. Jain, Anil. "The 5 V's of Big Data," IBM website, September 7, 2016. https://www.ibm.com/blogs/watson-health/the-5-vs-of-big-data/.

361. Lohr, Steve. "The Origins of 'Big Data': An Etymological Detective Story," *New York Times,* February 1, 2013. https://bits.blogs.nytimes.com/2013/02/01/the-origins-of-big-data-an-etymological-detective-story/.

362. Mayer-Schonberger, Viktor, and Kenneth Cukier. *Big Data: The Essential Guide to Work, Life and Learning in the Age of Insight.* Houghton Mifflin Harcourt, 2013.

363. Neal, David T., Wendy Wood, and Jeffrey Quinn. "Habits—A Repeat Performance," Duke University website, ca. 2006. https://dornsife.usc.edu/assets/sites/545/docs/Wendy_Wood_Research_Articles/Habits/Neal.Wood.Quinn.2006_Habits_a_repeat_performance.pdf.

364. Rudin, Cynthia. "Algorithms and Justice: Scrapping the 'Black Box'," *The Crime Report,* January 26, 2018. https://thecrimereport.org/2018/01/26/algorithms-and-justice-scrapping-the-black-box/.

365. Siegel, Eric. *Predictive Analytics: The Power to Predict Who Will Click, Buy, Lie, or Die.* Wiley, 2016. First published 2013.

366. Vigen, Tyler. *Spurious Correlations.* Hachette Books, 2015.

367. "VNI Global Fixed and Mobile Internet Traffic Forecasts," CISCO website, ca. 2017. https://www.cisco.com/c/en/us/solutions/collateral/service-provider/visual-networking-index-vni/vni-hyperconnectivity-wp.html.

CHAPTER 19—CYBERSECURITY AND QUANTUM COMPUTING

368. Armerding, Taylor. "The 18 Biggest Data Breaches of the 21st Century," *CSO,* December 20, 2018. https://www.csoonline.com/article/2130877/data-breach/the-biggest-data-breaches-of-the-21st-century.html.

369. Benioff, Paul. "The Computer as a Physical System: A Microscopic Quantum Mechanical Hamiltonian Model of Computers as Represented by Turing Machines," *Journal of Statistical Physics,* May 1980, Volume 22, Issue 5, pp. 563–591. https://link.springer.com/article/10.1007/BF01011339.

370. "Centre for Cybersecurity," *World Economic Forum,* ca. 2019. https://www.weforum.org/centre-for-cybersecurity/.

371. "Chapter 1—The global outlook for cyber security," in *Australia's Cyber Security Sector Competitiveness Plan 2019*, Australian Cyber Security Growth Network website, 2019. https://www.austcyber.com/resources/sector-competitiveness-plan/chapter1.

372. Cukier, Michel. "Study: Hackers Attack Every 39 Seconds," University of Maryland website, February 9, 2007. https://eng.umd.edu/news/story/study-hackers-attack-every-39-seconds.

373. "Cyber Crime Costs Global Economy $445 Billion A Year: Report," https://www.reuters.com/article/us-cybersecurity-mcafee-csis/cyber-crime-costs-global-economy-445-billion-a-year-report-idUSKBN0EK0SV20140609.

374. Fernandez, Joseph John. "Richard Feynman and the Birth of Quantum Computing," *Medium*, January 4, 2018. https://medium.com/quantum1net/richard-feynman-and-the-birth-of-quantum-computing-6fe4a0f5fcc7.

375. "Global Information Security Survey," Ernst & Young website, ca. 2019. https://www.ey.com/en_gl/giss.

376. "Guglielmo Marconi—Biographical," The Nobel Prize website, September 28, 2019. https://www.nobelprize.org/prizes/physics/1909/marconi/biographical/.

377. "Hack," Online Etymology Dictionary, n.d. https://www.etymonline.com/word/hack.

378. Hong, Sungook. *Wireless.* The MIT Press, 2010. First published 2001.

379. Howse, Derek. *Nevil Maskelyne: The Seaman's Astronomer.* Cambridge University Press, 1989.

380. "(ISC)^2 Finds the Cybersecurity Workforce Needs to Grow 145% to Close Skills Gap and Better Defend Organizations Worldwide," (ISC)^2 website, November 6, 2019. https://www.isc2.org/News-and-Events/Press-Room/Posts/2019/11/06/ISC2-Finds-the-Cybersecurity-Workforce-Needs-to-Grow--145.

381. Marks, Paul. "Dot-dash-diss: The Gentleman Hacker's 1903 Lulz," *NewScientist,* December 20, 2011. https://www.newscientist.com/article/mg21228440-700-dot-dash-diss-the-gentleman-hackers-1903-lulz/.

382. Wikipedia, s.v. "McEliece cryptosystem." https://en.wikipedia.org/wiki/McEliece_cryptosystem.

383. NASDAQ. "Cybersecurity: Industry Report & Investment Case," NASDAQ website, January 18, 2017. http://business.nasdaq.com/marketinsite/2017/Cybersecurity-Industry-Report-Investment-Case.html.

384. Porolli, Matías. "Cybercrime black markets: Dark web services and their prices," welivesecurity website, January 31, 2019. https://www.welivesecurity.com/2019/01/31/cybercrime-black-markets-dark-web-services-and-prices/.

385. Wikipedia, s.v. "Post-quantum cryptography." https://en.wikipedia.org/wiki/Post-quantum_cryptography.

386. Press, Gil. "Cybersecurity by The Numbers: Market Estimates, Forecasts, and Surveys," *Forbes,* March 15, 2018. https://www.forbes.com/sites/gilpress/2018/03/15/cybersecurity-by-the-numbers-market-estimates-forecasts-and-surveys/#7ce70ea612c4.

387. Wikipedia, s.v. "Quantum mechanics." https://en.wikipedia.org/wiki/Quantum_mechanics.

388. Wikipedia, s.v. "Quantum superposition." https://en.wikipedia.org/wiki/Quantum_superposition.

389. Shannon, C.E. "A Mathematical Theory of Communication," *The Bell System Technical Journal,* Vol. 27, pp. 379–423, 623–656, July, October 1948. http://www.math.harvard.edu/~ctm/home/text/others/shannon/entropy/entropy.pdf.

390. Smith, Tony. "Hacker Jailed for Revenge Sewage Attacks," *The Register,* October 31, 2001. https://www.theregister.co.uk/2001/10/31/hacker_jailed_for_revenge_sewage.

391. Teare, Gené. "Almost $10B Invested in Privacy and Security Companies in 2019," crunchbase news website, January 29, 2020. https://news.crunchbase.com/news/almost-10b-invested-in-privacy-and-security-companies-in-2019/.

392. Wikipedia, s.v. "Tech Model Railroad Club." https://en.wikipedia.org/wiki/Tech_Model_Railroad_Club.

393. Wikipedia, s.v. "Timeline of quantum computing." https://en.wikipedia.org/wiki/Timeline_of_quantum_computing.

394. Yagoda, Ben. "A Short Story of 'Hack'," *New Yorker*, March 6, 2014. https://www.newyorker.com/tech/elements/a-short-history-of-hack.

CHAPTER 20—PAST FUTURE

395. "American Check Their Phones 96 Times a Day," Cision PR Newswire website, November 21, 2019. https://www.prnewswire.com/news-releases/americans-check-their-phones-96-times-a-day-300962643.html.

396. Arbesman, Samuel. *Overcomplicated: Technology at the Limits of Comprehension.* Portfolio, 2016.

397. Bridle, James. *New Dark Age: Technology and the End of the Future.* Verso, 2018.

398. Desjardins, Jeff. "How Long Does It Take to Hit 50 Million Users?" *VisualCapitalist,* June 8, 2018, Chart of the Week. https://www.visualcapitalist.com/how-long-does-it-take-to-hit-50-million-users/.

399. Wikipedia, s.v. "Eroom's law." https://en.wikipedia.org/wiki/Eroom%27s_law.

400. Freedman, David H. "Lies, Damned Lies, and Medical Science," *The Atlantic,* November 2010, Technology. https://www.theatlantic.com/magazine/archive/2010/11/lies-damned-lies-and-medical-science/308269/.

401. Geekdom. "The History of Home Movie Entertainment," ReelRundown website, January 13, 2014. https://reelrundown.com/film-industry/The-History-Of-Home-Movie-Entertainment.

402. Wikipedia, s.v. "George Owen Squier." https://en.wikipedia.org/wiki/George_Owen_Squier.

403. Horrigan, John B. "Information Overload," Pew Research Center Internet & Technology website, December 7, 2016. https://www.pewresearch.org/internet/2016/12/07/information-overload/.

404. Ioannidis, John P.A. "Why Most Published Research Findings Are False," *PLOS Medicine*, August 2005. http://robotics.cs.tamu.edu/RSS2015NegativeResults/pmed.0020124.pdf.

405. Wikipedia, s.v. "List of the most popular websites." https://en.wikipedia.org/wiki/List_of_most_popular_websites.

406. Love, Dylan. "Laserdisc Sucked: The Evolution of Watching Stuff at Home," *Business Insider*, September 21, 2011. https://www.businessinsider.com/evolution-home-video-2011-9.

407. Morris, Chris. "Blu-Ray Struggles in the Streaming Age," *Fortune*, January 8, 2016, Tech. http://fortune.com/2016/01/08/blu-ray-struggles-in-the-streaming-age/.

408. Wikipedia, s.v. "Muzak." https://en.wikipedia.org/wiki/Muzak.

409. Wikipedia, s.v. "Philip K. Dick." https://en.wikipedia.org/wiki/Philip_K._Dick.

410. Robertson, Adi. "James Bridle on Why Technology Is Creating a New Dark Age," *The Verge*, July 16, 2018. https://www.theverge.com/2018/7/16/17564174/james-bridle-new-dark-age-book-computational-thinking-interview.

411. Scannell, Jack W., Alex Blanckley, Helen Boldon, and Brian Warrington. "Diagnosing the Decline in Pharmaceutical R&D Efficiency," *Nature Reviews Drug Discovery*, Vol. 11, pp 191–200. https://www.nature.com/articles/nrd3681.

412. Soulo, Tim. "Top Google Searches (as of July 2019)," Ahrefs (blog), July 1, 2019. https://ahrefs.com/blog/top-google-searches/.

413. "The Rise of Elevator Muzak Began with This Michigan Inventor," Michigan Radio website, September 13, 2017. http://www.michiganradio.org/post/rise-elevator-muzak-began-michigan-inventor.

414. Vosoughi, Soroush, Deb Roy, and Sinan Aral. "The spread of true and false news online," *Science*, March 9, 2018. https://science.sciencemag.org/content/359/6380/1146.

415. Wikipedia, s.v. "War of the currents." https://en.wikipedia.org/wiki/War_of_the_currents.

416. Wikipedia, s.v. "Year 2038 problem." https://en.wikipedia.org/wiki/Year_2038_problem.

ABOUT THE AUTHOR

GUY PERELMUTER is the founder of GRIDS Capital (www.gridscapital. com), a deep tech venture capital firm focusing on artificial intelligence, robotics, life sciences, and technological infrastructure.

He earned his Bachelor of Science in Computer Engineering in 1994 and a Master of Science in Electrical Engineering in 1996, both from the Pontifical Catholic University of Rio de Janeiro in Brazil. He specialized in computer vision techniques using artificial intelligence. In 1997 he was one of the winners of the Brazil Young Scientist Award for the implementation of his solution for the production of texts in Braille using dot matrix printers, and he later went on to develop risk analysis systems for financial markets.

His book *Present Future* was recognized by the Brazilian Book Chamber as the Best Science Book of 2020 in the 62nd edition of the annual Jabuti Prize.

Guy enjoys playing the piano, shooting hoops, watching science fiction, and reading (a lot). He currently lives in Brazil with his wife and their son and daughter.

CPSIA information can be obtained
at www.ICGtesting.com
Printed in the USA
LVHW090013270421
685671LV00015BA/145